DIAMONDS OF THE NORTH

DIAMONDS
OF THE NORTH

A CONCISE HISTORY OF BASEBALL IN CANADA

WILLIAM HUMBER

Toronto   New York   Oxford
Oxford University Press
1995

Oxford University Press
70 Wynford Drive, Don Mills, Ontario M3C 1J9

Oxford   New York
Athens   Auckland   Bangkok   Bombay
Calcutta   Cape Town   Dar es Salaam   Delhi
Florence   Hong Kong   Istanbul   Karachi
Kuala Lumpur   Madrid   Melbourne
Mexico City   Nairobi   Paris   Singapore
Taipei   Tokyo   Toronto

and associated companies in
Berlin   Ibadan

Oxford is a trademark of Oxford University Press

Canadian Cataloguing in Publication Data

Humber, William, 1949–
    Diamonds of the North : a concise history of baseball
in Canada

Includes bibliographical references and index.
ISBN 0-19-541039-4

1. Baseball – Canada – History.   2. Baseball players –
Canada.   I. Title.

GV863.15.A1H86 1995   796.357'0971   C95-930507-6

Front cover: *Canada versus the United States. An international baseball game at Lord's Cricket Ground in England during the First World
War.* Public Archives of Canada PA 022669

Back cover, from top to bottom: *The Maple Leaf Base Ball Club of
Guelph, Ontario, 1874.* Courtesy Guelph Civic Museum and the
Sleeman Breweries. *A women's baseball team, from Bocabec in
Charlotte County, New Brunswick, c. 1985.* Provincial Archives of
New Brunswick P71/27. *Turn-of-the-century baseball scene from
Bowmanville, Ontario, c. 1900.* Courtesy Bowmanville Museum. *The
Coloured Diamond Baseball Team of Halifax, Nova Scotia.*
Photography Collection, Public Archives of Nova Scotia. *Joe Carter.*
Courtesy Howard Starkman, Toronto Blue Jays Baseball Club.

Copyright © Oxford University Press 1995
1 2 3 4 - 98 97 96 95
This book is printed on permanent (acid-free) paper ∞
Printed in Canada

Contents

	Introduction	1
Chapter One	Canada in the Country of Baseball	3
Chapter Two	Early Baseball in Canada	15
Chapter Three	19th Century Ontario Baseball	23
Chapter Four	Maritime Baseball	41
Chapter Five	Western Canada	
	Part One, Alberta Baseball	62
	Part Two, Saskatchewan Baseball	75
	Part Three, Baseball in Manitoba	82
Chapter Six	British Columbia	93
Chapter Seven	Quebec	106
Chapter Eight	Baseball in One Canadian Community	127
Chapter Nine	Black Baseball in Canada	139
Chapter Ten	Owners, Organizers, and Players in Ontario Baseball	150
Chapter Eleven	Toronto Blue Jays	
	Part One, World Series Champions	168
	Part Two, The Jays do it again	177
Appendix A	Canadian Big League Players	183
Appendix B	Canadian Women in All-American Girls Professional Baseball League, 1943–54	198
Appendix C	Canadian Teams in Organized Baseball	201
Appendix D	Asahi Roster (1914-1941)	212
Appendix E	Expos and Blue Jays Historical Records	215
	Bibliography	217
	Index	224

This book is dedicated to all my friends in the two continuing education subjects I have taught every year since 1979 at Seneca College. One makes the winter go faster and the other on the road in Cooperstown is the best weekend of the summer.

INTRODUCTION

Future historians will note at least two moments in Canada's past when an aversion to Toronto was overcome by the victory of that city's baseball team in the World Series. These were moments that electrified an entire nation.

Ken Burns' 1994 nine-part PBS broadcast, *Baseball*, suggested that with the Toronto Blue Jays' 1992 World Series victory the sport had become truly international. Nothing, however, could be further from the truth. For when Canadians from coast to coast poured out on the streets to celebrate the first of Joe Carter's famous leaps, they were also proclaiming the sport's magnificent heritage in Canada.

This book has its genesis in an innocuous reference in Irving Leitner's book, *Diamond in the Rough*, to a game of primitive baseball played in the small community of Beachville, Ontario in 1838. Not only was this one year before the date claimed for the game's invention in Cooperstown, New York, but it suggested a heritage of baseball in Canada far more extensive than most would expect.

Historians Bob Barney and Nancy Bouchier have confirmed the essential details of the Beachville game as baseball's predecessor, townball, which in turn descended from English rounders and had its roots in fertility rites dating back many thousands of years and celebrating spring's arrival.

In *Cheering for the Home Team* (1983) I began the process of uncovering baseball's forgotten roots in all parts of Canada. The game has been played, in one form or another, throughout the country since the early 19th century. I have come to suspect as well

that, given baseball's late arrival as a major league sport in Canada—Montreal in 1969 and Toronto in 1977—Canadians are somewhat more open to studies that focus on the history of baseball outside the top level of the game.

In many ways the story of baseball in Canada is that of the lost tribe, as it were, of American baseball history. Japan's game was largely separate from that of North America, and baseball in the Caribbean has few well-developed connections to the North American game beyond its export of talented ballplayers in the post-war period. Baseball in Canada, on the other hand, has been intimately connected to the evolution of the game in the United States. Canadian teams have been a part of organized baseball's structure beginning with London's and Guelph's entry into the International Association in 1877; exhibition contests date back to the time of the American Civil War. Canadian-born ballplayers, executives, umpires and even owners have been a part of the American baseball structure continuously since 1871.

Periodically, correspondence from American researchers seeking information on a Canadian relative of Honus Wagner's wife, or details about an obscure ballplayer who played seven games in Calgary prior to World War One, remind me of our mutual history. Yet the story of baseball in Canada is far more extensive than its links to baseball in the United States. Its independent quality is evident in its play among all ages and social groups, and both sexes, in all parts of the country. If baseball were not already the national pastime of the United States I suspect we would long ago have declared it ours.

In examining the history of baseball's popularity in the everyday life of Canadians and, so I hope, shedding light on that life, I was able to draw upon a marvellous variety of written and verbal sources to whom personal and professional recognition is extended in the bibliography. I also note the tremendous support provided by libraries including that in Bowmanville, Ontario, Seneca College, the Metro Reference Library, the Public Archives of Canada and the Baseball Hall of Fame and Museum in Cooperstown, New York.

Ultimately baseball's attraction is a very personal thing that I share with my family, Cathie, Bradley, Darryl, and Karen at great events like the 1992 and 1993 World Series in Toronto, with fellow workers who share a baseball enthusiasm, among them Jan Richards, Steve Wilson, Peter Crawford, Eleanor Sutton, Susan Scobie, Vitra Garcia, Donna Bailey, Tony Tilly, Wayne Norrison, as well as a host of friends from my two annual baseball courses at Seneca College including Dave Crichton, Bill Maxwell, Butch Crotin, Anthony Kalamut, Rosemary Branson, George Auerbach, Russell Field, Gord Kirke, Wendy Wilson, Corinne Chevalier, Steve Stohn, Ian Hamilton, Veronica Trevithick, Wayne Vekteris, Al Laurie, John Carriere and his gang, Bill Murtagh and his crew, Bob Gilson, Larry Wood and

so many others, and finally the bantam age baseball players on the team I coach in the summer and particularly our pennant-winning team, Bowmanville Glass and Mirror, in 1994.

William Humber is a writer, teacher, and organizer of events associated with baseball. His books on the topic include *Cheering for the Home Team* (1983), *Let's Play Ball: Inside the Perfect Game* (1989), and *The Baseball Book and Trophy* (1993). He wrote the sections on baseball for both editions of the Canadian Encyclopedia (1985 and 1988), was subject specialist for the Royal Ontario Museum's 1989 exhibit on baseball, and is the only Canadian to have served on the Board of Directors of the Society for American Baseball Research. In a classroom setting at Seneca College he teaches an annual mid-winter Baseball Spring Training for Fans and each year leads a summertime trip to the Baseball Hall of Fame in Cooperstown, New York. Bill is a regular contributor on baseball matters to radio and television shows throughout Canada. His article on Canada's baseball history appeared in the 1993 World Series program. He has also written books on bicycling, soccer, and his hometown of Bowmanville, Ontario. He is a Chair in the Faculty of Continuing Education at Seneca College in Toronto.

CHAPTER ONE

Canada in the Country of Baseball

There are supreme moments in Canada's infatuation with baseball such as those following Joe Carter's ninth-inning home run in the sixth game of the 1993 World Series that gave the Toronto Blue Jays their second successive baseball title. The streets of the country overflowed with unrestricted joy from Vancouver's Robson Street to downtown Saskatoon where local police had to be called out; from Pond Inlet on the northern edge of Baffin Island where the baseball season begins as it ends in -10 degree weather and Inuit children play baseball during recess imagining themselves to be Roberto Alomar, to the lower end of Ontario's several thousand kilometre long Highway 11; from the outport of Butlerville, Newfoundland where they cheered the next thing to a native son and even to many Québécois households in the baseball strongholds of Trois-Rivières and St-Hyacinthe—though the pleasure was tempered by memory of the Montreal Expos' failures of the early 1980s.

At other times, however, baseball pains Canadians. They wrestle awkwardly with their support for the game. It is a kind of guilt associated with a love for what is after all the national game of their neighbours to the south. The Japanese feel no such remorse, having imbued baseball with characteristics peculiar to their culture. Cubans, who have reason to be suspect of anything so symbolically American, revel in their ability to beat the Yanks at annual amateur tournaments.

In Canada successive journalists from Goldwin Smith, the country's 19th-century man of letters, to author Lawrence Martin in his 1993 book *Pledge of Allegiance* on the Americanization of Canada in the Mulroney years, have bemoaned the intrusive nature of American culture on Canadian life and used baseball as their symbol. 'A national game cannot fail to exercise a great influence on character,' Smith wrote in August 1880. 'What is the national game of Canada to be—Cricket, Lacrosse, or Base Ball? . . . Base Ball has now gained a strong hold upon this Continent, and all the circumstances are in its favour. It can be played through the spring, summer, and fall; it does not require much of a lawn; and what is a greater advantage still, it is quickly played, so that the game is commonly finished in an afternoon. . . . The loyal Englishman [in Canada] who regards with pensive regret the adoption of a Yankee game may console himself with the thought that Cricket and Base Ball have apparently evolved out of the same infantine British sport.'

Over a hundred years later, writing in *The Toronto Star* (25 April 1993) about the symbolism of US President George Bush attending a ballgame in Toronto's SkyDome in 1990, Lawrence Martin said, 'The fantastic success of baseball north of the US border was yet another sign of the times. Major league ball now enjoyed charter membership in English Canada's new continentalist culture. In sports, as in music, film, and books, it was a culture less conscious of borders.'

This attitude is at least somewhat peculiar in that baseball was adopted in Canada while still in its primitive stage and long before it could be an agent of cultural imperialism. This is a claim that no other country outside the United States can make. The English, after all, who were responsible for bringing rounders to the new world had consigned this particular bat and ball sport to the rubbish heap of failed games, leaving it to be played by small children while adults got on with the more serious game of cricket. Canadians from Victoria, British Columbia through to Halifax, Nova Scotia, were effectively regional participants in the game's evolution, playing what might be thought of as experimental forms of baseball from which the modern game emerged.

Until at least 1860 the game flourished as a significant regional variation in southwestern Ontario alongside other local interpretations in Philadelphia, New York, and New England. Even after the popular New York game was adopted in Canada around the time of the American Civil War, until 1876 slight rule variations still existed. This is a game with a significant heritage in Canada though one undeniably connected at a very early stage to events in the United States. Far from being victims of American taste in the matter of baseball Canadians were at the very least junior partners in the game's development. Such involvement was significant in that it showed an emerging Canadian interest in new, North American traditions rather than inherited British ones.

There was baseball of a type in Canada at least 30 years before Canadian Confederation in 1867. The absence of national direction to develop the sport in Canada was compounded by the powerful regional character of the game as reflected in local rules in the United States and Canada. Baseball in Canada has always had a pronounced north-south orientation. Maritimers played against east-coasters from Massachusetts and Maine, Québécois moved among the baseball circles of Vermont and New Hampshire and only occasionally eastern Ontario, Ontarians competed in Michigan, Ohio, Pennsylvania, and New York states. Manitobans joined leagues in Minnesota and the Dakotas, Alberta and Saskatchewan welcomed itinerant ball clubs from the American Plains states, and in British Columbia games were played along the American Pacific Coast.

Lacking a national identity competition sought symbols other than those for national honours. In 1863 funds were solicited for a Silver Ball competition, based on an American model, to be contended for by teams in southwestern Ontario, by far the most advanced baseball region in any of the areas that now make up Canada. Likewise baseball in Canada was following the commercial path taken by the game in the United States. Culturally and politically this connection had its roots in the 1837 Upper and Lower Canadian Rebellions which borrowed democratic ideas and active help from American sources (Upper Canada Rebellion leader William Lyon Mackenzie fled to the United States after the uprising failed). From its earliest stages baseball in Canada as in the United States was played by a wide variety of social classes and supported by a small business entrepreneurial class. As such it was part of the populist commercial entertainment favoured by the American public and gradually adopted by working-class Canadians. More privileged classes in Canada, argue Richard Gruneau and David Whitson (1993), 'were drawn primarily towards the model of British "gentlemanly" amateurism'.

American-based associations representing the different regional variations of baseball had been established in the United States in the 1850s, but Canadian teams were much too primitive to take part in these activities. By the time baseball in Canada had sufficiently matured, Americans were at war with themselves and baseball was slowly assuming an American identity following the triumph of the New York game. Baseball was no longer a folk game of quaint local initiative, and Canadians took tentative steps to form a Canadian Association of Base Ball Players at the time of the Provincial Exhibition in Hamilton in late September 1864. It was an early recognition of an emerging Canadian national identity at a time when the Fathers of Confederation were still debating the form of the new country. Significantly this Canadian Association was made up solely of teams from southern Ontario, a nomenclature at least excusable in 1864 as the territory of Canada West and East included only present-day Ontario and Quebec and there was little evidence of baseball activity in Canada East.

The Silver Ball championship lasted through 1876. In the 1860s it involved towns and cities from Woodstock and Ingersoll in the west to the village of Newcastle in the east, and reached its farthest extension in 1873 with teams from Ottawa and Kingston. A distinct Canadian nationalism did not infuse these competitions as it would do so later in lacrosse, hockey and football. Baseball's allegiance to commercial interests had already made it the perfect vehicle for touting the economic prospects of newly emerging urban centres. For men like Guelph brewer George Sleeman a baseball team was a means of promoting his business and his town in both the immediate hinterland and potential markets many miles away. The regional nature of competition also reflected the available means of transportation.

The first significant international event, that of the great Base Ball Tournament in Detroit in August 1867, included teams from Michigan and Ontario. Little notice was taken of national origins until the Victorias of Ingersoll were asked to allow a rival team to use a substitute for a late-arriving player, 'the Canadians refusing to allow Porter to play in his place'. There was little bitterness in the comment and no further reference to their Canadian background though the Detroit *Free Press* did say that 'the hard-fisted laboring men' of Ingersoll should have participated in a higher class of the tournament.

William Bryce's 78-page 1876 *Canadian Base Ball Guide* was the first significant Canadian publication on the sport. It is notable for the total absence of any reference to baseball outside Ontario, a lapse inexcusable by that year though well in keeping with a popular misconception of Ontarians that they spoke for the rest of the country. Bryce was a publisher, book-seller, and sporting goods distributor whose Ontario Game Emporium was based on Richmond Street in London, Ontario. He was entirely conscious of baseball's national character, noting 'The organization of a Canadian Association of Base Ball Players [at Toronto's Walker House on 7 April 1876] and the adoption of a special constitution and by-laws, playing rules, and championship code, have made the publication of a Canadian Base Ball Guide a necessity.' He acknowledges 'the want of authentic records of the game in Canada', but gives not the shade of a hint that this want may extend beyond Ontario's borders. His historical knowledge is sparse: he comments that 'Within the past six years Base Ball has made rapid strides in public favor in Canada, and in the western and northern portions of Ontario, especially, it has to a great extent displaced Cricket and Lacrosse as a favorite summer out-door recreation.' His patriotism was aroused by the exploits of Guelph's Maple Leafs in the mid 1870s who 'did not even rest content with Canadian conquests, but carrying the Maple Leaf across the border gained decisive victories over the best of the American so-called amateurs. Professionals, too, have had in one or two instances to lower their banner to the Canadian champions.'

The gradual importation of American professionals into Canadian baseball ranks in the early 1870s culminating in Kingston securing 'almost an entire nine from over the border', created little public concern. Indeed, by 1876 Bryce says 'The impetus given to base ball by the greatly improved style of play resulting from the introduction of foreign talent, is manifesting itself in the increased patronage bestowed upon the game by the public in all parts of Canada.' Concern about the importing of pro ballplayers almost always reflected the increased costs they brought to local entrepreneurs and the unsavoury association of such players with known gamblers. George Sleeman was notably guided by his pocketbook, alternately cursing imports one year and hiring them the next.

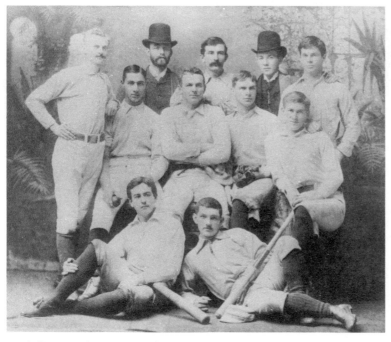

Baseball team at the University of Toronto, c. 1887. Metropolitan Toronto Reference Library, T 13353

The import issue was not unique to Canada. Residents of many American cities also decried, though obviously not in national terms, the replacement of the locally born ballplayer representing his town by the itinerant mercenary who played today and was gone tomorrow. In the final analysis the majority of fans, or kranks as they were then known, supported the system because it improved the quality of local play. More of them then attended this popular form of entertainment and this in turn encouraged entrepreneurs to hire even better players to improve their chances of winning. It was a simple equation, well understood by William Bryce. He realized in 1876 that, in the same year as the National League's formation in the United States, a Canadian response was natural. 'Base Ball', he said, 'has now reached such a stage of perfection in Canada that its leading clubs are able to cope successfully with the best of the same class in the United States.'

Significantly for many Canadians in Ontario and, in a less pronounced but nevertheless evolving sense elsewhere, baseball was becoming part of everyday life. Casual play was common. The obituary of Joseph Gibson of Ingersoll, one of that town's 'hard-fisted' players, recorded that some of his hap-piest moments were playing baseball with his six boys in the park adjacent to their home. Likewise Tom Gillean, born in London in 1855, played the game so well as a youth that he was one of the few Canadians to play on the champion London Tecumsehs of 1877, a participation, said *The London Free Press,* that 'pleased those who believe imports are not necessary, when good locals are available'.

The decision of the Tecumsehs not to join the National League for the 1878 season probably had little discernible impact on baseball's future in Canada; their tenure, in those turbulent times for baseball teams, likely would have been short-lived. Still it was an opportunity that would not come again until the Montreal Expos joined the National League for the 1969 season. As the quality of professional play improved so did the desire of entrepreneurs to limit this success to a few specialized and larger centres. The reserve clause introduced in the early 1880s allowed National League teams to restrict the free movement of players and in time made possible the establishment of baseball's Minor Leagues, classified from Triple "A" to "D" by levels of play and to some extent by urban size. Smaller centres unable to compete with the major leagues for baseball talent were

Ottawa ball club of 1898 in highland tartan

protected in a structure of major and minor leagues termed 'Organized Baseball'. The territorial markets and player contracts of local entrepreneurs were respected, allowing many of them to prosper. As this 'organized' structure slowly took shape in the last two decades of the 19th century it created a mixture of resentment and resignation among hold-outs. The nominally independent and Ontario-based Canadian League of 1885 collapsed after one year when Toronto and Hamilton joined the American-run International League, one acknowledging its subservience to the major leagues. In the process William Southam's *Hamilton Spectator* bluntly declared, 'Baseball is a pure matter of business and can't survive on the friendship of Guelph.' Baseball's civic connections were now in a more complex commercial entertainment structure. Hamilton and Toronto might declare their economic dominance over a surrounding region but they were also being swallowed within a structure subservient to the interests of baseball's major leagues.

The present-day structure of organized baseball was realized in 1902 when the National Association of Professional Baseball Leagues was established to oversee the minor leagues' relationship with the major leagues. Under a National Agreement in place before 1902, Spalding's 1900 *Guide* noted, 'the National League of the United States [was] the power that rules the whole professional base ball business of the great republic.' Cal Davis, president of the Canadian League, made the case for Canadian subservience.

> The question of the relation of the minor base ball leagues to the National Board, and of the National Agreement to the game of base ball, is one which opens a very wide field for discussion. . . . from the standpoint of the Canadian League . . . the league has had a continuous existence almost since the old International League disbanded eight or nine years ago, but that only during the years 1897, 1898, and 1899 was the league a member of the National Agreement. It is my personal experience, and that of every manager in the league, that the condition of the game in Canada has been so vastly superior

under the protection of the National Agreement to what it was as an independent organization that the league would sooner pass out of existence than return to the old ranks again.

Davis then mentioned the adjudication by the National Board on matters regarding disputes between his league and others of higher gradings, between clubs of different leagues, and between individual players and clubs.

Davis's response was that of a pragmatic businessman anxious to survive in the complex world of organized baseball but it shouldn't surprise that a few years later Samuel Moffett's report *The Americanization of Canada* slammed these apparent intrusions into Canadian affairs. Moffett's comments on baseball, however, were too narrowly concerned with the effects on British traditions and missed the larger implications implicit in Davis's rationalization. 'It is not a trivial matter,' Moffett wrote, 'that baseball is becoming the national game of Canada instead of cricket. It has very deep significance, as has the fact that the native game of lacrosse is not able to hold its own against the southern intruder.' Among Moffett's telling points were the Ontario Legislature's cutting short its session so that the provincial premier could throw out the first ball of Toronto's 1905 Eastern League season, and the Mayor of St Thomas's declaration of a holiday after three o'clock on the first day of the Western Ontario League season. 'The Canadian newspapers,' Moffett fulminated, 'print fuller telegraphic accounts of the great baseball contests of the National, the American and the Eastern Leagues than they do of the proceedings of the British Parliament,' and furthermore, 'The American baseball language, which would be entirely unintelligible to an English reader, is fully acclimated in the Canadian press.'

One of the first attempts to counter baseball's north-south axis by forming a truly nationwide baseball organization appeared in Spalding's 1914 *Canadian Base Ball Guide*. 'From the east to the west the game has advanced to that extent that one can readily expect to see within a very short period of time a Canadian League that will in all probability be

> Toronto no longer goes wild over bicycle races, although it howls itself to death at a baseball game.
>
> *Cycling* magazine, 1900

composed of a circuit embracing the following cities: London, Hamilton, Ottawa, Toronto, Montreal, Quebec, Saint John, and Halifax.' Only excessive mileage, the writer argued, prevented extension of such a league to western Canada. The vision conformed to similar developments tying Canadian hockey and football together from coast to coast. In baseball's case however it was not realized, partly because the First World War intervened. More importantly, such a league in Spalding's vision would have to exist within the structure of organized baseball, and that being the case the north-south orientation made better business sense.

Historian Alan Metcalfe argues that by 1914 'baseball was truly Canada's national sport. No other sports were played across the country and exhibited such steady and sometimes spectacular growth.' Cricket had declined to an inconsequential status; its clubs were often made up of older men like William Southam and Thomas Goldie, who played baseball as youths, but opted in their later years for cricket's social status. Field Lacrosse had grown from a minor status in the last century after George Beers promoted it as a proper national game for the new Canadian nation. In making his case for lacrosse Beers acknowledged he had to work fast since baseball, 'that American game', was gaining in popularity. A game

can grow only so far on patriotic claims and by the time of the First World War lacrosse also was in decline. Its downfall is often attributed to its brutality but a more convincing argument was the game's failure to usurp baseball's successful integration into so many aspects of Canadian life. Baseball was played by men and women of all ages and social classes, a claim lacrosse could not make. As the skill level of baseball improved and threatened to chase away the casual player the game proved incredibly adaptable. Softball was gradually introduced between the First and Second World Wars and it expanded the participant base for baseball-type games. The slow pitch variation would have similar effect in the 1970s and 80s. Though purists might object, the two variations have much in common with baseball of the 1860s period. Lacrosse also tried to adapt by introducing the game of box lacrosse in indoor arenas but in the process killed the field game as an outdoor summer sport.

Some sense of baseball's popularity in this period is the level of newspaper coverage. In 1915 it was the most reported sport in Halifax (37.6%), Montreal (18.6%), Winnipeg (21.2%), and Edmonton (32%). In Toronto its 23.5% coverage made it the second most reported sport. A study by Evelyn Waters for the period 1926 through 1935 ranked baseball and hockey as almost even in newspaper coverage from coast to

Baseball team of the 4th Canadian Division Signal Company, July 1918, somewhere in France; courtesy Clay Marston

coast. Baseball ranked first in Toronto and Vancouver, and second in Montreal, to hockey while both sports were about even in Halifax and Winnipeg.

The game had completely surpassed all other summer games and was arguably the most popular game in Canada. Hockey was still growing and in the absence of large numbers of indoor ice rinks was limited in many parts of Canada by weather conditions ranging from January thaws to chinook winds. Yet even as baseball reached levels of popular support it was dogged by its American connection. In a 1920 essay Archibald MacMechan attacked the Americanization of Canadian sport: 'Our native game, lacrosse, is dead. Cricket, which flourishes in Australia, is here a sickly exotic. But baseball is everywhere.' One of the characters in New Brunswick writer George Clarke's 1926 novel *The Best One Thing* expressed the contempt of at least some Canadian critics: '. . . baseball was demoralising the youth of the land. If it was only cricket, now; a gentleman's game; a moulder of character. . . '. The cultural complaint had at least some justification from a social context. In the case of Dick Brookins, an alleged black player in the Western Canada League in 1910, league officials cowardly sought guidance from the National Association on whether he should be allowed to play. At least in this case it could be argued that as a member of the Association the League was obliged to consult American officials. The same could not be said of an incident recalled by London newspaperman and historian Les Bronson. 'A baseball team protested the use of a negro player, by, I believe, Ingersoll in the Ontario Baseball Association playoffs in the early twenties. It would be after

> One undated news-clipping (probably about 1930) tells of a July Sunday at King's Wharf 10 years earlier, when the boys played in Sullivan's cow pasture, dispensing with an umpire and with only a shadowy knowlege of the rules: 'It was a long drawn-out affair... the lads didn't quit til 'twas milking time... just a whale of a good time by all on the one day they had off in a week. They don't play anymore. Some of the younger bucks got married and moved away, some of the older ones passed on, and O'Leary's house burned down (1926) with all the baseball equipment in it...'
>
> p. 251, *A History of the Township of Emily in the County of Victoria* by Howard Pammett (1974)

Landis was named commissioner for I believe a letter was sent to his office to secure a ruling whether the player was eligible or not.' As a Canadian amateur organization with no affiliation to organized baseball the Ontario association's resort to an American authority was a damning indictment of the game's subservience to foreign control.

This example may explain at least in part the game's somewhat troubled state in the late twenties when for several years only one Canadian-born player appeared on a major league roster and Toronto was the only city with a team in organized baseball. The only comparable period in the 20th century was the late 1960s.

Through several generations of play baseball had become a highly specialized game requiring superior bat and ball skills. Increasingly major league ballplayers were being drawn from parts of the United States where the game could be played year round. In the first decade of the 20th century there were over 500 ballplayers from the states of New York, Pennsylvania, and Ohio (which shared much in common with nearby Canadian centres) as opposed to fewer than 60 from Texas and California. The three northeastern American states retained and even slightly increased their major league participation between 1910 and 1919 (due partially to the Federal League's brief existence which for a time created an additional 200 major league jobs). California and Texas however contributed over 150 players. In the twenties contribution from the three northeastern states had fallen to just over 350 while Texas and California surpassed 200. The New York, Pennsylvania and Ohio total of major leaguers fell below 300 in the thirties,

Canada versus the United States. An international baseball game at Lord's Cricket Ground in England during the First World War. Public Archives of Canada PA 022669

Canadian women's armed forces team, the Eager Beavers, and the officers of the Regina Rifles play baseball at the end of the Second World War. The Netherlands, 31 August 1945. Public Archives of Canada, PA 116314

barely ahead of the number from the two southern states. Adding to this territorial shift in ballplayer birthplaces was the challenge of other sports. Many skilled athletes in Canada were now focusing their major league ambitions on hockey.

The period between the wars also saw the gradual spread of what were then simpler bat and ball games like softball. Softball's rapid spread drew potential players away from baseball. The earliest forms of softball were in fact indoor, winter variations of baseball played in the 1880s. Following the 1885 season members of Hamilton's professional teams, the Clippers and Primroses, had played baseball on roller skates at the Royal Roller Rink at Main and Catharine. (Among the Primroses was J.H. Barnfather who later played baseball in Winnipeg.) At the turn of the century the Toronto Garrison Officers' Indoor Baseball League met at the old University Avenue Armories; Sir Henry Pellatt was one of the directors. Experiments such as these in many North American cities led to refinements like the underhand pitching motion, a larger, less lively ball, and shorter distances between bases. These common sense responses to the playing environment were eventually taken outdoors. Softball's development was a sign that ordinary folks were simply trying to find an accessible means of connecting to the

1967 Canadian Pan American Games Baseball Team: back row (L-R) Joe Zeman (manager), Al Robertson, Ron Smith, Larry Smith, Tom McKenzie, Cam Hurst, Glennis Scott; middle row Gladwyn Scott (assistant coach), Irv Doerksen, Ken Ewasiuk, Ed Tanner, Bob Hunter, Lane Jackson, Don Summer (statistician); front row Gerry MacKay (coach), Bob McKillop, Cliff Seafoot, Maurice Oakes, Gene Corey, Phil Dorion, Larry Bachiu. Missing from picture — Ross Stone, Ron Stead, and John Elias; courtesy David Shury

sport. As in time softball developed all of the professionalism and skill refinement of baseball, non-pros evolved the game of slow pitch.

Softball's popularity opened the doors to greater participation in bat and ball games. 'Women,' says Colin Howell, 'could be more easily integrated into [softball] as players,' and women were attracted to softball in large numbers. From Toronto's Sunnyside to the playgrounds of Saskatchewan, a generation of women developed bat and ball skills which some of them later used in the All-American Girls Professional Baseball League. Within the native communities of north-eastern North America softball was even more attractive than baseball because gloves were not always required.

At the national level, baseball's interests were largely subsumed within the Amateur Athletic Union of Canada. Given that so much of the game was controlled by local amateur sandlot and commercial interests the game's lack of authority on the national stage was not a serious hindrance to the game's growth. In April 1937, however, the Canadian Amateur Hockey Association broke away from the Amateur Union, and was followed by secessionists

from a number of other sports including baseball. Though the effect wasn't immediate, the way was paved for the eventual development of a meaningful national organization.

Though the 1940s and 50s were in fact boom years for minor league baseball in Canada, at the grass-roots level the game was faced with significant competition from other forms of recreation including automotive travel, television, summer cottages, and other spectator sports. Significantly all of these pastimes signalled an increasing privatization of life in which the simple public joy of playing or watching baseball with friends was giving way to more individual personal pursuits. Writing about baseball in Nova Scotia between 1946 and 1972, author Burton Russell described the common refrain of the day: 'Some people with myopic vision or short memories tend to pass off baseball. They point out that the game is dying and that drag racing and skiing have capture the wallets, and the fancy, of the young people.' While disputing their claim Russell nevertheless gave it credence in his observation that 'Spectator interest in Nova Scotia senior baseball has declined drastically during the last several years. How unlike the tremendous enthusiasm displayed for the game back in the thirties, forties, and early fifties when thousands of fans poured through the gates to witness senior baseball at its best.'

Interest in senior baseball in Nova Scotia reached such a low point that in 1968 the province's senior baseball league folded, and senior ball suffered the same fate in Cape Breton. Only one senior club, playing in a junior league during the season, registered for playoff action. Baseball's long minor league reign in Toronto had ended the year before. Even in the United States the sport had been replaced by football as the dominant national pastime. Theories accounting for baseball's decline referred to its slower pace, though such arguments failed to account for the game's internal logic and the role of pace in creating necessary dramatic moments. More reasoned points of view suggested that the game's delicate balance between offence and defence had become weighted in the pitcher's favour, citing the 1968 major league season in which Denny McLain of the Detroit Tigers won over 30 games and . Bob Gibson of the St Louis Cardinals had a scintillating earned run average of 1.12 runs per game.

The response at the major league level was to lower the pitcher's mound from 15 to ten inches and create a smaller strike zone. The most effective remedy, however, was that sure cure for too much good pitching—expansion. Among the new teams brought into the majors was the National League's first foreign entry, the Montreal Expos. In creating conditions for baseball's reinvigoration on the field, major league owners also encouraged the gradual resurgence of baseball interest throughout Canada.

The chief cause of baseball's apparent decline in the sixties had little if anything to do with the game itself. The post-war North American economy was expanding exponentially as virtually all of its competitors from Japan to Germany to Great Britain were forced to rebuild their industrial infrastructure. As the dominant North American summer sport baseball was the first entertainment to take advantage of this affluence. By the early fifties the game had dangerously overexpanded; its hundreds of markets would be incapable of supporting professional baseball at the slightest downturn. The same could be said of much senior baseball like Nova Scotia's Halifax and District League which had a quasi semi-professional character. As the 1950s progressed the public began to use its new wealth to indulge in a

variety of other private amusements. Not surprisingly baseball was forced to retrench but in doing so appeared to be in decline. In an affluent society appearance counts for much, and many supporters turned away from baseball rather than be identified with a public loser. Compounding this problem was the baby boomers' rejection in the 1960s of much of their parents' culture, of which there was no better symbol than baseball.

However, the game's problems had the positive benefit of forcing its amateur organizers in Canada to take remedial action. Attempts to give the sport some form of national direction dated back to the formation of the Canadian Baseball Association in 1864, again in 1876, and sporadically thereafter. These were largely regional initiatives, and short-lived. Baseball was a game dominated by commercial interests whose decisions were based on business considerations not those of national organization. The Canadian Amateur Baseball Association had been formed in 1919 but even at the amateur level regional interests dominated. In response to threats to the game's very existence an organization that today is the Canadian Federation of Amateur Baseball (CFAB) was incorporated in 1964.

The formative years of the CFAB were difficult ones. David Shury of North Battleford, Saskatchewan, secretary of the organization was given responsibility for putting together a national team to represent Canada in the 1967 Pan-American Games in Winnipeg. Recalls Shury, 'Bob Lacoursiere of Saskatoon and I had taken over earlier that year and had inherited a very sorry organization. It was broke, the provinces were in open revolt and the national organization was in difficulties with Ottawa. The Federal Government refused to let us have any funds with which to form a team.'

Nevertheless under coach Gerry MacKay from Manitoba a team was cobbled together which included a few players with past professional connections in Quebec and Ontario. Though reinstated as an amateur, pitcher Ron Stead was known to Toronto sportswriters from his try-outs with the Toronto Maple Leafs a decade before. At the time he had been profiled in a CBC documentary. The Olympic code which applied to the Pan Am games strictly forbade the participation of anyone who had past professional connections and news of the Canadian transgression broke during the competition. Shury later blamed Toronto sportswriters eager for a story. None of the other teams complained, he said, because they were all in the same boat. Four Canadians were sent home and the demoralized team did not win a game until their final match with undefeated Cuba. Canada's 10-9 win forced Cuba to play the United States for the gold and the Americans scored an unexpected upset.

The glimmerings of national direction for the game of baseball outside the structure of organized baseball almost immediately suffered another blow. In 1968 the federal government commissioned a study of amateur sport. The resulting report applauded the role of sports in fostering national unity and recommended federal involvement based on a sports priority standing. Priority one sports would receive favourable funding and services but as David Shury learned in early 1970 baseball was not among them, primarily because it lacked Olympic standing. Baseball officials launched an immediate public relations campaign that included support from New Brunswick Premier Louis Robichaud and former Prime Minister Lester Pearson, then honorary President of the new Montreal Expos. When the Minister of National Health and Welfare John Munro made his announcement of designated sports in April 1970, 21 sports had been designated as priority one and baseball was among them. As Shury says, '[Munro] had also mentioned that he did not realize so many people were interested in baseball.'

The 1970s were kinder to baseball. The Montreal Expos began play in a converted amateur stadium, Jarry Park, in 1969. The endearing charm of that park, the team's roster of colourful players (a characterization that in fairness is applied to all expansion teams), and their national television exposure had immediate impact. In Montreal in 1968 there were only 222 amateur teams at all levels of play. That number jumped to 430 within a year. Torontonians in a burst of civic pride were doubly miffed by the award of the 1976 Olympic Games to their rival. Led by Metro Toronto Chairman Paul Godfrey and a number of commercial interests, most notably Labatt Breweries, Toronto's pursuit of a major league baseball franchise began in earnest. Baseball's long decline through the 1960s was ending. The game was simply too popular and too historically engrained to fade away and its stabilization corresponded to an era of new stars and great moments—like Carlton Fisk's home run in the sixth game of the 1975 World Series, which is often mentioned as being symbolic of the game's more adventurous offensive style. Writers—from *The New Yorker's* Roger Angell to the novelist stylings in Roger Kahn's remembrances of the old Brooklyn Dodgers, Jim Bouton's off-the-wall observations from a player's perspective, and the statistical imagination of Bill James—were presenting the game in a fresh manner to a new generation of fans and renewing the faith for the jaded former follower. The Toronto Blue Jays who entered the American League in 1977 were an immediate sensation, drawing nearly two million fans including over 40,000 who attended their inaugural match in, what else, a Canadian snowstorm.

Figures released by the CFAB indicated that in 1979 there were over 160,000 registered baseball players in Canada including 50,000 in Quebec and 31,000 in British Columbia. Little League, Babe Ruth, and Connie Mack associations were not included but CFAB Executive Director Paul Lavigne suggested they accounted for another 25,000 players. CFAB figures showed only 12,000 registrants in Ontario, a number that actually shrank to under 11,000 in 1981. In fact CFAB figures also showed decline in Quebec (down to 43,000) and British Columbia (29,035). With the exception of Prince Edward Island all Maritime provinces had shown growth. The most feasible explanation for this apparent decline in registrations is a natural stabilization in participation after the initial excitement surrounding major league baseball's arrival in Canada. By 1988 the game's popularity and participation rates were less subject to the first blush of expansion. In that year participation including Little League, Babe Ruth leagues, and regular amateur baseball affiliated with CFAB, totalled over 200,000.

Between 1990 and 1993 participation exploded at all levels except the senior (slipping from 6,700 to 5,800) where it is possible many were drawn into house league teams, softball, or slow pitch. In CFAB's affiliated recorded ranks at the youngest level, Mosquito baseball grew from 42,000 in 1990 to 74,000 a year later, and 96,000 by 1993. Pee Wee grew from 25,000 in 1990 to 41,000 three years later, Bantam grew slowly from 17,000 in the same period to nearly 25,000 by 1993, while Midget slumped from 10,000 after 1990 but recovered with 12,000 participants in 1993. Junior baseball accounted for only 3,200 in 1990 and 4,600 in 1993. (Many of these figures are misleadingly low because they do not include local house leagues in which

Typical small-town Canadian amateur baseball team: the 1994 Bowmanville Glass and Mirror Bantam house league team for 14- and 15-year-olds; courtesy William Humber

most children play.) This participation was rewarded in 1991 when Canada's national youth team (18 and under) won the country's first-ever amateur Gold Medal at the World Youth Championship in Brandon, Manitoba. One of the team's pitchers, Joe Young from Fort McMurray, Alberta, was drafted by the Toronto Blue Jays in 1993.

Orleans in the east end of Ottawa's National Capital Region offers an example of baseball's explosive growth in Canada. Orleans is a relatively new community but one that sought to put down roots by developing a youth sports program. Formed in 1983 with about 150 players, the Orleans Little League grew within a decade to 730 players, a maximum dictated by the limited number of ball diamonds. Players range in age from 8 to 18 and are drawn from an area that includes the eastern part of Gloucester and the western part of Cumberland township. Coaching and umpiring have improved over time and all-star teams compete as far away as Montreal and northern New York state. The impact of the Expos, the Blue Jays, and in 1993 the Triple A Ottawa Lynx has caused many

children to rethink their decision to devote all their dreams to a hockey career. Junior hockey players with the Ottawa 67s keep their options open by playing baseball in the Orleans system.

The number of participants in fast and slow pitch softball is astounding. In 1988 Robert Barney noted that perhaps as many as two million Canadians played some version of the underhand game; just over 10 per cent were registered in some form of league play leading to recognized championships. These participants were split almost evenly between males and females. Barney notes by way of contrast that hockey including its connected forms of girls' ringette and, largely, men's ball hockey accounts for about half a million players.

In 1994 the Canadian Parliament confirmed hockey as Canada's winter game and lacrosse as the country's summer sport. If participation and national enthusiasm, such as that which followed Joe Carter's home run mean anything, there can be little doubt that the real national pastime of Canada has always been baseball.

CHAPTER TWO

Early Baseball in Canada

Among the legion of baseball scribes there is no mention of Ely Playter but his diary entry of 13 April 1803 reads 'I went to town . . . walked out and joined a number of men jumping and playing ball, perceived a Mr Joseph Randall (a farmer from near Newmarket) to be the most active.' Playter operated a tavern in York, a community of fewer than 700 residents, and the game he played may be a primitive form of either baseball or football. Supporting the theory that this was an early bat and ball exercise are the rules of the English game of rounders in which players leaped and jumped to avoid a ball tossed at them. In rounders as in the game that grew out of it, American townball, an offensive player was declared out if hit by a tossed ball while between bases, a practice known as soaking.

The community of York would be incorporated as the city of Toronto in 1834. Yet 30 years after Playter's ambiguous reference bat and ball games other than cricket were still not accorded formal public recognition because of the low esteem in which they were held. A century before in England the landed aristocracy's economic dominance had been supplanted by a new industrial class of entrepreneurs. Stranded on their rural estates they passed the time by, among other means, playing games and the one they approvingly adopted was suited to their wide expanses of flat green lawns. We now know it as the game of cricket, but until then it was but one of a wonderful variety of bat and ball games such as cat, knur and spell, trapball, rounders, and stoolball played by both sexes and all ages in English village

sites in the middle ages. These games in turn descended from bat and ball games played as fertility rites in the darkest recesses of European history. By selecting cricket as the mature adults' game, the English upper class effectively condemned all other bat and ball games to a lower status, and in keeping with that judgement those games withered and often disappeared. In the case of rounders its failed estate in England restricted its play there to young children. To this day the game is associated with the recreation of young ladies at the beach on bank holidays.

The term rounders derives from the French 'ronde' which described the circuits of the base in the French game of theque or tec. The word 'ronde' and the games associated with it came to England from France, where it met a variety of Scandinavian games such as 'northern spell' and eventually became rounders. Played on a diamond with a base at each corner, rounders required a 'feeder' to toss the ball in an arc to a batter, who was allowed three refusals. This game came to America with immigrants from the southeastern counties of England where the two-word term 'base ball' was used in place of rounders. It was a game perfectly suited to idle recreation at New England village meetings where it was played in a style more in keeping with its folk tradition in medieval English villages than its later degraded state following the adoption of cricket as England's dominant bat and ball game. In America it was a game in search of a uniformity and adult respectability it lacked in England. New Englanders called it 'townball', and no one rebuked them for playing with

Unknown player photographed by John Cox of Kings Road, Hamilton, Canada West; c. 1868 uniform similiar to that worn by cricketers, before the distinctive baseball outfit developed; courtesy Mark Rucker

Joe Hornung of the London Tecumsehs baseball team, 1877; courtesy Mark Rucker

enthusiasm and purpose what some perceived to be a children's game. It was not played very well and when cricket was finally established in America in the early 19th century the best bat and ball players were found on the cricket pitches.

Baseball in Upper Canada

There are only a limited number of references to early bat and ball games in North America prior to 1840. Even cricket has a restricted heritage in the new world, limited in this period primarily to the garrison and élite schools such as Upper Canada College in present-day Toronto. Authenticated accounts of bat and ball games other than cricket are sparse and, like Playter's, somewhat ambiguous. They had nothing in common with today's baseball, which is characterized by commerce and civic pride. They were games integrated into the fabric of everyday life—common activities that received little recorded attention. Those accounts that do exist indicate the game spread far

afield, though it is difficult to judge the depth of its play. Was it rare, sporadic, or daily?

The most famous account of an early baseball type game in Canada is that in Beachville, Ontario near Woodstock played on 4 June 1838. The description of this game appeared almost 50 years later in a letter in the 5 May 1886 edition of *The Sporting Life* of Philadelphia. Dr Adam Ford of Denver, Colorado, who was born on a farm in Zorra Township of Oxford County probably in 1831, wrote an account of the game following his exile to the American west after having been accused of poisoning a temperance leader in the doctor's office in St Marys, Ontario.

Adam Ford's memory was extraordinary. Though he was only seven at the time of the Beachville game he remembered player names of whom we know that Nathaniel McNames was 24, Daniel Karn 19, Harry Karn 17, and William Dodge 15. Dodge's father had served on the British side in the War of 1812. The June 4 game marked a traditional celebration of King George III's birthday (Ford errs in listing George IV) known as Military Muster Day. Most significantly this date was reserved for celebration, as Ford noted 'the rebellion of 1837 had been closed by the victory of the government over the rebels.' The infield was square, the baselines 24 yards apart and five bases were used. The teams represented the small settlement of Beachville, and the Zorras from the township of Zorra and North Oxford.

The ball was made of double and twisted woollen yarn, and was covered with calfskin. The club (bat being a term used in cricket, Ford says) was generally made of the best cedar blocked out with an axe and finished on a shaving horse. Ford described the rounders and American townball rule of soaking, and recalled that there were fair and foul balls. Innings could range from six to nine and players per side from seven to twelve depending on how many showed up. Most significantly he recalled that a battalion of Scots volunteers, off to fight the remnants of the previous year's rebellion, watched the game. Colonel A.W. Light had written Captain Gibson that month stating 'We of Woodstock, Ingersoll, and Zorra are coming to you as soon as we can muster our men . . . the Lieutenant Governor at Niagara has taken nearly all the rascals who attacked the Lancers and I trust we shall do the same with the scamps near Zorra . . . '. Significantly, Colonel Light lived a mile and a half from a Beachville wagon-maker, Cornelius Cunningham, originally from Vermont. Cunningham was a secret leader of the rebels; he was captured following the battle of Windsor on 4 December 1838 and executed on 4 February 1839.

The Beachville game is particularly notable because it occurs in the midst of the only significant rebellion in the history of present-day Ontario. Better economic conditions and the promise of a simple, cheaper government appealed to Upper Canada reformers like William Lyon Mackenzie who had visited the United States in 1829. The costly and speculative nature of land disbursement and patronage in Upper Canada controlled by an educated Tory clique, dubbed the Family Compact, left the region ill-equipped to develop its own resources. Too much land was unsettled and decision-making was encumbered by a quasi-aristocracy out of touch with the new business class (the same class that would eventually support baseball). Residents of southwestern Ontario looked to Ohio, Illinois, and Indiana for leadership in issues of concern such as non-sectarian common schools, roads and bridges, solemnization

of marriage, the franchise and elections, tenure of office, and freedom of religion. And for the common person it was the tavern or inn where new ideas were discussed by Americans passing through from Niagara or Detroit. As likely as not one of those new things discussed was the emerging game of baseball.

Emigration from the United States had begun at the end of the 18th century and continued to the War of 1812. Two groups of former Americans, one loyal to the crown and the other convinced the area would one day join the United States, swelled the census rolls of southwestern Ontario and competed for influence in the region.

On the one hand were those with few American sympathies. The Treaty of Paris, signed in 1783 following the American Revolutionary War, was supposed to prohibit reprisals against those Americans who had remained loyal to the crown. It failed in this purpose and fleeing loyalists complained of atrocities. During his tenure in Upper Canada in the 1790s Lieutenant-Governor Simcoe attempted to reinforce the region's British ties by reserving London as the colony's future capital—a decision overruled by Sir Guy Carleton who instead chose the site of what is now Toronto.

Another of Simcoe's decisions had more lasting significance. He issued wholesale settlement invitations to American citizens. In Beachville the children of many of these ex-Americans such as 16-year-old William Dodge, Warner Dygert, and Chris Karn all fought on the British side in the War of 1812. Their descendants were to play in the 1838 game. Some former Americans only grudgingly swore allegiance to the King in return for free land. Near Beachville in the village of Delaware Ebenezer Allen, the village's founder, and Andrew Westbrook, a leading citizen, supported the invading Americans. Westbrook led many raids into the region and in an August 1814 attack on the Beachville district burned the mill and carried off several leading citizens including Captain John Caroll, son of the district's original settler. Caroll was killed by a rescue party who mistook him for Westbrook.

A powerful anti-American sentiment spread throughout Upper Canada following the war but American influence continued to penetrate the region. Americans were among London's earliest entrepreneurs. They included George Goodhue, postmaster and retailer whose business provided everything from groceries to the rope used for public hangings, and tanners Simeon Morrill, London's first mayor, and Ellis Hyman, later prominent in the London, Huron and Bruce Railway which linked London with its northern hinterland. The fact that Americans brought their informal bat and ball games with them should hardly surprise.

The rebellion had probably been doomed to failure when the colony's governor Sir Francis Bond Head blamed Upper Canada's political squabbles on Americans and so was able to attract to the loyalist side even those with limited sympathies for the Family Compact. A leading Tory newspaper, *The Patriot,* perhaps paraphrasing Bond Head, declared on 15 July 1836 'A cricketer as a matter of course detests democracy and is staunch in allegiance to his king'—which brings us to the true significance of the Beachville game. The early days of 1838 were dangerous times in southwestern Ontario. Though many in the region of Beachville would have sided with American popular opinion on a variety of issues, and a significant portion of them had American roots, this was not the time to be celebrating an American lifestyle. Why then would the citizens of Beachville publicly play such a quintessentially American sport in front of troops off to fight rebels whose cause was supported by American ideals and arms?

The reason quite simply is that Beachville citizens would have found nothing peculiar in their primitive baseball play of 1838 because as yet the game had no national identity. Though the term 'national game' was first found in the 22 August 1855 minutes of New York's Knickerbocker Base Ball Club, the game did not have such general standing in the United States until after the Civil War.

This non-aligned game which made its way into Upper Canada sometime in the early 19th century was adopted by Canadians to meet their own needs and interests. In its 4 August 1860 issue, the *New York Clipper* acknowledged the game's unique features in Canada, noting that 'the game played in Canada differs somewhat from the New York game, the ball being thrown instead of pitched and an inning is not concluded until all are out, there are also 11 players on each side.' Today's baseball is not so much an American invention as a game refined and tested in many forms throughout North America. In its transition from primitive play, baseball developed regional variations. Townball for instance eventually became known as the Massachusetts Game in honour of its origins, and other styles included the Philadelphia and New York games.

Canadians were among the regional partners in the establishment of the modern game of baseball and contributed to the sport's evolution, standardization, and modernization. Their influence was regional but at least as significant as that of nearby ballplayers in western New York and other states to the south. The eventual triumph of the New York game which, though formulated in the mid 1840s, did not surpass its rivals until the Civil War, should be looked on only as part of a necessary evolution of rules allowing for conformity and play between various regions of North America.

The Beachville game and Ford's account suffers not by any lack of detail but for the opposite reason. It is almost too good to be true. Historians Nancy Bouchier and Robert Barney who have studied the game in great detail speculate that it may mark a crossover from a more primitive game to that eventually codified in New York City by Alexander Cartwright among others. Such a direct evolution from one form to another, however, seems less likely than that from a variety of regional experiments one ultimately reigned supreme. Ford's account offers its greatest veracity in the boundaries of its detail—the time, the general format, and the players. It becomes more questionable and subject to his later interpretation of events as he explains the rules. His absolute description of distances between bases seems somewhat preposterous. Ford was after all a seven- or eight-year-old boy at the time.

Towards the end of his account Ford recalls returning to Beachville from his studies at McGill University perhaps as early as 1849 but more likely at some time in the early 1850s. The local residents were then apparently aware of the new New York rules which expressly forbade throwing the ball at a runner, allowed for the use of a harder ball, and limited the number of players on the field to nine. Ford's account is significant for several reasons. This may be our lone description of baseball play in Canada between 1842 and 1854. Ford was residing in Quebec, but gives no indication that either the New York game or baseball in general was played there at the time. Secondly the formal recorded arrival of the New York game in Ontario does not occur until 1859. The two decades after 1838 were clearly a period of great and largely unrecorded transition in the game. The New York game had to establish roots in its own region before conquering others. Communication with Upper Canada was sporadic. By stage and steamboat it was a four-and-a-half-day trip from Windsor to Oswego, New York on the south shore of Lake Ontario and several more days from there to New York City.

Ford suggests that the New York game violated local understanding of what constituted baseball. 'We went out,' he says, 'and played it one day. The next day we felt as if we had been on an overland trip to the moon.' The railway was still a few years away and with it the quicker means of transportation and contact that would reinforce these early contacts with the New York game.

Organized teams first appeared in Hamilton in 1854 and London in 1855 and they played a game having much in common with the Massachusetts game. More significantly Canadian adults were acknowledging by their participation that what was once a folk game had become a serious adult activity in North America. Canadians like Americans were gradually opting for baseball over cricket. Unlike cricket, whose well-codified rules frustrated North Americans, baseball was still a game adaptable to local circumstance. They wanted to play a game that could be completed in several hours to conform to their busy work schedules and one that accordingly allowed for a greater exchange between offence and defence. Baseball was as well a game that still welcomed the unskilled, unlike cricket where a generation of skilled English cricketing immigrants intimidated native Americans still learning the game. Given the choice many North Americans opted for what was then the easier game to play, baseball. The course that baseball in Canada set itself on with the formation of the Hamilton Maple Leafs in 1854 is but the first step towards the Toronto Blue Jays World Series victories of 1992 and 1993.

The end of the Canadian type game and adoption of the New York rules occurred in the late 1850s and early 1860s. Before reviewing those events it is worth taking a step back and considering the multitude of other encounters Canadians had with baseball prior to its eventual public adoption during the American Civil War. It is a story with few if any connections between the various parts of Canada. Individual accounts appear to suggest that the old European bat and ball games arrived not only through the United States but sometimes directly from overseas. Combined these accounts provide a powerful understanding of how baseball sank vital and significant roots into the Canadian landscape prior to its adoption as America's national game in the latter part of the 19th century.

Rounders comes to British Columbia

Three thousand miles west of Beachville, Ontario, the British Columbia Archives holds the notes and comments of James Robert Anderson describing his school experiences in Victoria in the period roughly between 1848-51. Victoria is located on the southern tip of Vancouver Island about 100 kilometres from Vancouver. It was chosen for settlement in 1843 by James Douglas, the chief factor of the Hudson's Bay Company's Pacific Coast headquarters, and remained a small community of a few hundred people until the Fraser Gold Rush of 1858. Writing of a time about eight years after his birth in 1841, Anderson says,

> Our amusements consisted of marbles, cricket, rounders, shinny, horse riding, fighting Indian boys, worrying Indian dogs, some surreptitious shooting with our antiquated flint lock muskets. . . . Captain Grant, late of the Scots Greys, God Bless him, was our patron as regards cricket, having presented us with a full set, which enabled us to indulge in the game which was usually played on the ground just where the Burns Memorial now stands. Balls for rounders, the game now called baseball, and for shinny we constructed of hair covered with dressed deer hide.

These games were likely imported directly from England. The term rounders never had currency in the sporting language of the United States and its childhood play in Victoria conformed to its position in England. Victoria's essentially British character at this time and mention of cricket's prominence would seem to corroborate this conclusion.

' Bat ' in the Red River Settlement

At a mid point between Victoria and Beachville was the west's first significant European settlement in Red River, a colony founded in 1812 at the junction of the Red and Assiniboine Rivers in present-day Manitoba on land granted by the Hudson's Bay Company to Lord Selkirk. Schofield's 1913 history of Manitoba describes life in the 1830s. 'During the summer months the people were too busy for much amusement; but the gun and the fishing rod furnished sport for the holidays, and "bat", a game of ball in which leagues and professional players had no part, gave recreation during the long evenings.'

The game was undoubtedly a variant of bat and

ball games popular throughout Europe and was brought directly to the Red River from that source. Manitoba historian Wilson Green speculated in a 1981 correspondence on its actual play.

> I am inclined to think (from the name) it may have been similar to a rough and primitive game we played as small (and not so small boys), requiring only sticks or clubs and something in the shape of a ball, sometimes even a frozen horsedropping (as in road hockey). All but one player (no set number under say eight) dug a shallow depression in which he kept his 'bat' until getting a swipe at the dead or rolling ball in any direction. While one was doing this, another could slip his stick into the vacated base. In this way the odd player could occupy any vacant 'hole' he found. The game got underway by one lad heaving the bundle of 'bats' as far as possible, excepting his own. The player who was unable to return to his (or any) depression was required to keep the ball in play until finding an opening himself. We called this game 'pig'. It was based on skullduggery, not sportsmanship and required no equipment. The game invariably ended in one or more fights.

This play would seem to be connected to European bat and ball rituals reaching back thousands of years in which holes for burying a bat or ball symbolized the planting of new seed in the springtime. The Red River game appears to be a folk game imported directly from Europe to a relatively isolated settlement in what was then the Northwest Territories.

Other Ontario encounters with early baseball

The region around Beachville has many accounts of early baseball type games besides the one described by Ford. They are games of either direct British import or American influence. A series of articles written around 1900 for the *Woodstock Daily Express* described life in the 1830s in the village of Brighton, a small hamlet around what is now the intersection of Vansittart Avenue and Ingersoll Street in Woodstock. The author called the game baseball though this appears to be the application of a later nomenclature.

> Baseball was an evening practice of the young men after the day's work. The game was popular, match-

es were frequent with the villagers on the court house green, and the play exciting. The Pascoes, Gunns, the Egans, the Budds, and the Dunns were leading players. Imitation was strong in the juveniles. Every little fellow must have his bat and ball and many an old sock, and many an old rubber was ripped up, used up, and cut up, and wound up, and sewed up in sheepskin or, when they had the coppers, in calfskin by friend 'Benjie'. The higher the bound the better the ball.

A hard ball is inconsistent with the general use in both rounders and American townball of a soft ball thrown at a player between bases. This suggests the Brighton game may have been a relatively primitive one with few or no bases, emphasizing throwing and hitting. The players, says the writer, were British emigrants. 'They landed on the shore on their own responsibility; they landed in this district on their own resources and independently of each other.' Of the 21 families, 14 were English Episcopalians and Methodists. They were former military officers, carpenters, shoemakers and labourers and, perhaps most significantly, their direct route from the British Isles suggests that the game they played was an English folk version of baseball and not one that had passed through any American refinement.

Likewise in nearby London there is mention of the old English game of rounders. Interviewed by the *London Free Press* in 1911, 86-year-old William Peters recalled that as a child on Ridout Street in the late 1830s or early 1840s he played rounders on the old courthouse square. There was lots of fun, he recalled, adding 'the boys used to be pretty good shots with the ball'—a reference to the rounders rule in which the ball was thrown at the runner.

Quebec's first baseball game

In Quebec, historian Donald Guay in his *Introduction à l'histoire des sports au Québec* suggests a potential broadening of the base of bat and ball influences. 'This team sport,' he says, 'which is popular in the United States is a simple American adaption of cricket or the English games of "rounders", "Goal Ball", "Feeder" and "Round Ball".' Guay noted that a writer in Quebec City's *Le Soleil* in 1899 wondered if the modern version was really an old game of ball played long ago in Normandy and Brittany.

The original report in *Le Soleil* only notes the

similarity between modern baseball and ancient bat and ball games like La Grande Theque played in the north of France. The 1899 writer implies but offers no evidence that these old French games were at some point introduced into New France. Recording such apparently trivial matters as simple bat and ball play would be rare, and such games if they did cross the ocean would likely have declined in form and content to become the simplified play of children. Still it is not completely absurd to suggest that residual memory of such games could have survived, ultimately influencing young French Canadians in the late 19th century.

There is however one major reference to early baseball in Quebec in the notes Robert Sellar used in the preparation of his *History of the County of Huntingdon and Seigneuries of Chateauguay and Beauharnois,* printed in 1888. Born in Scotland in 1841 and briefly employed in his early twenties as a compositor for George Brown's Toronto *Globe,* Sellar was induced by a local committee of the Liberal Party to start a newspaper (*The Canadian Gleaner*) in Huntingdon, Quebec in 1863. Huntingdon forms part of the eleven Quebec townships south of the St Lawrence River and east of the Richelieu, an area known as Les cantons de l'Estrie. The area hugged the American border and its population was made up largely of descendants of United Empire Loyalists. Sellar took a particular interest in the lives of early settlers and his oral histories survived at least until the early 1940s. They were reprinted by his son over a period of four or five years in the renamed *Huntingdon Gleaner*. Local historian Wayne McKell paraphrased the key comments from several issues that had appeared in May 1941:

In the 1830s the town (then village) of Huntingdon in southwestern Quebec was the home of two distinct inhabitants. One group were settlers from Ireland, Scotland and England. The other was the American faction—opportunists rather than loyalists. The two groups did not mix well. The Americans were mainly merchants and a group of them operated the Canadian side of the Ogdensburg, New York to Montreal stage line. On fine summer evenings during the mid 1830s, a group of the American faction were playing ball in Huntingdon. One Hazelton Moore, the stage driver, threw a bean ball at the batter, Fisher Ames. Ames took exception to this insult, rushed Moore, and

struck him on the head with his bat. The fight was stopped by the other players and Moore had his injury treated by Dr Shirriff (whose son later started Shirriff's Foods) and the two men made up. However Moore went into a coma later that night and died the following day. In a curious twist of jurisprudence, Ames was cleared at the inquest when Dr Shirriff suggested that Moore had died because his skull was too thin.

Samuel Graham, one of Sellar's witnesses, said the event happened in 1837, the year of the great Lower Canada Rebellion. 'Ames was a very passionate man,' McKell records, 'and the first blow might be excused on that ground, but he struck him twice, the second blow when he was lying insensible on the grass. The Americans (Crawford and the rest) bribed Moore's wife to stay away and her absence at the trial helped get Ames off.'

And for his part Dr Shirriff told Sellar, 'The stage made good time in summer for it was after it came up (from Chateauguay Basin, Quebec—a day's journey away) that Hazelton Moore was killed. He was driver and joined in the game of ball in front of the (Lewis and Ames) store. Something that Ames said or did provoked him and instead of throwing the ball to him, he (Moore) threw it at him, when Ames rushed towards him and struck him with the club on the head.'

The ball used in games of that day would not be hard in recognition of the soaking rule. Throwing at a batter, however, would be unusual and given the character of the players in this game a reprisal strike with a bat does not seem too farfetched. The absolute authenticity of the account requires additional corroborating details but it is consistent with the appearance of early forms of baseball in nearby parts of the United States. Ironically a book of children's amusements released in 1830 in New York warned its readers about the improper use of a club or bat: 'the writer of this,' says the book, 'knew a youngster who had his skull broke badly with one and it nearly cost him his life.'

Early Maritime baseball

American ties also played a role in the early appearances of bat and ball games in the Maritimes. Writer Joan Payzant notes that the commons of both Halifax and Dartmouth were used for baseball in the 1830s and 40s. She quotes from the 1838 diary of a Haligonian that 'baseball came in with the May flowers and did not last much longer.' Another reference occurs in the 1 July 1841 edition of Joseph Howe's newspaper *The Novascotian* and may be the first mention of a baseball type game in a Canadian newspaper. On the afternoon of St John's Day, 24 June 1841, 700 to 800 members of the St Mary's Total Abstinence Society of Halifax (founded in 1838) crossed the harbour on the steamer *Sir Charles Ogle* to the strains of a brass band. They marched to 'a beautifully situated field, half a mile from Dartmouth, very kindly allowed for the occasion by Mr Boggs. . . . Quadrille and Contra dances were got up on the green—and games of bat and ball, and such sports proceeded.' While the bat and ball game could have been cricket, it is unlikely as that game was already well known and would have been listed as such. The British background of the marchers suggests that this bat and ball play has an overseas connection.

Reinforcing that conclusion is the research of historian Colin Howell who has documented the playing of what was known as the 'old fashion' game among natives of Atlantic Canada up to the time of the Second World War. Its similarity to rounders suggests that it was introduced to the Mi'kmaq, Maliseet, and Penobscot peoples by Christian missionaries and teachers in the mid 19th century before rounders was completely supplanted by baseball as we know it. The ball was made of yarn or sponge rubber and the game required few skills, allowing all ages and sexes to play.

Baseball in Halifax's maritime rival, Saint John, New Brunswick, also appears around this time. A centennial essay by D.R. Jack on Saint John, published in 1883, says: 'It was a common practice with many of the leading merchants of Saint John to assemble each fine summer afternoon after the business of the day was over, on the north side of King Square, where a fine playground has [sic] been prepared, and engage in a game of cricket or baseball.' The practice was continued until about 1840. J.S. Knowles writing in 1911 says, 'Baseball . . . was a game strictly amateur and the club from about 1853 to 1860 was composed of business and professional men.' According to Brian Flood, well-known citizens of the day played the game at low tide on the head of the Courtenay Bay flats and at high tide played in a field opposite Kane's Corner. What little is known of the rules suggests it was a regional variation of American townball. Teams had nine players but the entire side had to be retired in an inning (of which there were only three), and the soaking rule was in force.

Games in Newfoundland

The established game of baseball does not appear in Newfoundland until the early 20th century, a factor related to its isolation and position as a British colony. Paul O'Neill's history of St John's, Newfoundland says that 'An *Evening Telegram* reporter sneered at it [baseball] as a glorified game of rounders as played in his schooldays.' Whether those schooldays were in England or Newfoundland is not said but the very awareness of rounders suggests at least the possibility of its play in Newfoundland by British emigrants.

Conclusion

It is clear then that by the time formal teams appear in southwestern Ontario in the mid 1850s the game had had a lengthy incubation though there is no clear pattern to the variety of baseball type games appearing in various parts of Canada. Those in Victoria in 1849, in the Red River in the 1830s, in Brighton and London, Ontario in the same period, in Dartmouth in 1841, among Atlantic Canadian natives and possibly in Newfoundland, point to a direct British influence. More American in character are those in Beachville, Huntingdon, Halifax and Saint John appearing between 1837 and 1840.

It was a time when informal bat and ball games were about to enter a period of greater formality characterized by team organization and written rules. This next stage received its greatest impetus in the United States, particularly in New York City. A model emerged in the organization of the New York Knickerbockers. Though borrowed from cricket it was a powerful means by which baseball took the great step forward to challenge and eventually surpass cricket as the dominant bat and ball game.

CHAPTER THREE

19th Century Ontario Baseball

On 19 January 1854 the streets of Hamilton were jammed in celebration of the completion of the Great Western Railway linking Niagara to Detroit. The Mayor of Rochester, a city which had protected participants in the 1837 Upper Canada Rebellion, marched arm in arm with Hamilton's chief magistrate Charles Magill. They were followed by Hamilton's various societies, along with local militiamen and members of the volunteer fire companies. The *Hamilton Spectator*'s Bob Smiley described them as 'dressed in uniform, and a finer, more athletic, and if the term was appropriate, a more soldierly looking body of men I have never witnessed.'

Parades were one of the era's more obvious expressions of civic pride, and the athletic young men who marched in them were early organizers of another expression—the baseball team. With the completion of the Great Western Railway, the repeal of the Navigation Acts in Britain in 1849 ending Canada's favourable trading position in the Empire, and the introduction of free trade with the United States in 1855 Canadians, particularly those in southwestern Ontario, were in close contact with their American neighbours.

Until that year bat and ball games in Canada were informal affairs more characteristic of medieval folk sport than the modern game with which we are familiar. Everything changed in 1854 and one factor was no doubt the improved communications associated with new railway construction. Formally organized clubs had existed in the 1820s using cricket as their example. The New York Knickerbockers of the

1840s and their rulemaker Alexander Cartwright are generally credited with being the first important team but their real significance lies in their willingness to admit that they were first and foremost baseball players. Until then mature adults had used the somewhat rustic game of baseball, with its links to child's play and inferior bat and ball skills, as a warm-up for the more serious adult game of cricket.

Other clubs and leading organizers were also prominent in the 1840s and 50s. Indeed a debt is owed to New Yorker Daniel Adams who headed an American convention in 1857 which confirmed nine innings rather than 21 runs as the game's conclusion; a year later Adams recommended the 90-foot distance between bases.

Canada's first team from Hamilton, Ontario

The best evidence suggests that Bill Shuttleworth (1833–1903), a Hamilton clerk, organized Canada's first formal team, the Young Canadians (later the Maple Leafs), who played on the grounds facing Central School between Bond and Beverley Streets. His fellow players were a cross-section of Hamilton's working people and at one time or another they included five clerks, three shoemakers, two turners, two labourers, as well as a coach manufacturer, broom manufacturer, saloon keeper, hatter, painter, hammerman, foreman, butcher, machinist, marble-cutter, brakesman, carriagemaker, boilermaker, tin-smith, horse-collar maker, teamster, watchmaker,

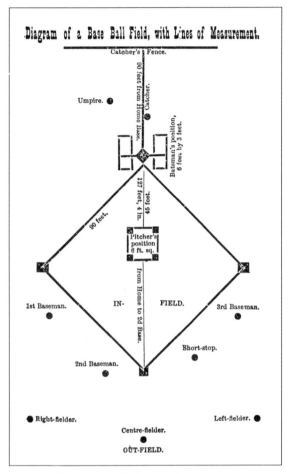

Diagram of the 1876 baseball diamond. Compare the 45-foot distance from the pitcher's position to the plate, with today's 60 feet 6 inches. Bryce's 1876 Canadian Base Ball Guide

cigarmaker, cabinetmaker, grocer, bookbinder, plasterer, tobacco maker, carpenter, sailor, and wool sorter. It was hardly surprising that Hamilton should be one of baseball's first strongholds in Ontario. It was a town experiencing all the boom, bust, and boosterism of similarly sized American cities. Its population had grown from 6,000 in 1846 to 16,000 ten years later. The Great Western Railway had given Hamilton a temporary advantage over Toronto in exploiting the business potential of western Ontario. A commercial depression, brought on by overspeculation in railroads after 1857, and Toronto's invasion of the western Ontario hinterland (with the completion of the Great Western Railway) would forever end Hamilton's ambition to control the economic future of the region. Nevertheless boosters continued to assert Hamilton's identity as the 'ambitious city' by

introducing band concerts in 1862 and an opera house in 1880.

Until the Civil War baseball was a regional game and one in transition from its informal, folk background. As a transitional game it was characterized by the adoption of semi-formalized though regionally differentiated rules and play largely within one's own club. The Canadian interpretation played in Southwestern Ontario can quite legitimately be considered the only experiment outside the United States on baseball's eventual form. It differed slightly from the Massachusetts game in its strict adherence to 11 men on the field as opposed to the Massachusetts rules which allowed 10 to 14. The Canadian game required that all 11 men be retired before the other team came to bat, as opposed to the requirement that only one man be retired in the Massachusetts game. Both games allowed the pitcher to throw the ball in the modern style, rather than underarm as in the New York rules. The first formal team in London, Ontario, organized in 1855, restricted its membership to 22 men so that they could be equally divided into two squads of 11 each. The additional positions in the Canadian game were a fourth baseman and a backstop behind the catcher.

This Canadian brand of baseball remained popular in Ontario until the end of the decade, as did the practice of playing games only against the members of one's club and not against other teams or towns. In New York City, clubs were now defining a new style

'The Arrival of the Bats', an 1884 street scene in Southern Ontario; courtesy Mark Rucker

of play increasingly remote from the easy-going fraternal character of earlier days. By the late 1850s, large unruly crowds were common. The Brooklyn Atlantics had their blue collar supporters; their rivals, the Excelsiors, depended on merchants and clerks. Gentlemanly ideas gave way to more vibrant encounters, as the sport became the focus for occupational groups. It was in places like saloons, cigar stores, pool halls, and bookmaking establishments, the general hang-outs of streetcorner society, that the modern form of baseball emerged. Perhaps not surprisingly therefore it was New York's rules which were slowly winning ascendancy among baseball players.

The New York rules are essentially the rules of the game today. They permitted three outs, thus maximizing individual opportunities to play. They forbade the "soaking" rule which hence allowed use of a harder ball that could be hit out of the infield. They limited the playing area to that bound by the 90 degree territory defined by third base, home, and first base and thus permitted teams to use only nine as opposed to eleven men on the field, thus offering increased opportunity to bat. The rules appealed to players and spectators alike because they encouraged a quick exchange between offence and defence. They were codified and, once universally adopted, not subject to regional interpretation. New York's general reputation even in those days as a cultural trend-setter further enhanced its power to direct baseball's future. Not incidentally the increasing opportunity for matches against other clubs, a factor less of rules than the aggressive interest of New York clubs in matching their skills against rivals, marks the development of baseball into a fully modern game.

As late as 1860 the Young Canadian Club of Woodstock were playing the eleven-a-side Canadian version of baseball. They had been organized, on the grounds by Reeves Street back of the post office, by Bill Shuttleworth's younger brother Jim. On this day they outlasted The Rough and Ready Club of Ingersoll 83–59. Both towns are within a short buggy ride of Beachville, where the first accredited bat and ball game had been played in 1838. And though there is no direct evidence, it seems reasonable to conclude that this version of the Canadian game between Woodstock and Ingersoll evolved from games such as those played at Beachville. One of Woodstock's first settlers was Zachariah Burtch, whose family had been represented in that 1838 game. Woodstock was the centre of a growing region.

The railroad ensured local prosperity and in the 1860s a cheesemaking industry was founded by Harvey Farrington, an immigrant from New York State's Herkimer County (ironically near the present site of the National Baseball Hall of Fame and Museum in Cooperstown). By 1867 there would be 35 factories and Ingersoll cheese was popular in Hamilton shops.

In 1859 teams from Hamilton and Toronto played the first recorded game in Canada using the New York rules. Perhaps not surprisingly given the standardization of play fostered by the New York rules this game also is the first mention of a game in Canada between teams representing rival cities.

The Hamilton Burlingtons' loss to the Niagaras of Buffalo in 1860 in the first ever international match sealed the fate of the regionally differentiated Canadian game. Buffalo had adopted the New York rules in 1857 and the pragmatic young merchants and businessmen of the Burlington club, which had been formed in 1855 with 50 members playing mid-week club matches on the grounds near Upper James Street and Robinson, willingly made the switch. The Hamilton Maple Leafs debated the merits of the new rules. Arthur Feast, a stonemason, supported Bill Shuttleworth's defence of the Canadian game. Charles Wood, a young innkeeper from the United States, won the day with his support of the New York game. A year later Wood convinced the Woodstock players to switch to the New York rules and the Canadian interpretation of baseball disappeared for all time. On May 31 the Young Canadians of Woodstock led by Wood played a picked nine from their town, winning 10-7 in the first match that had been played there using the New York rules. Of the players in that match, Jim Shuttleworth, Robert McWhinnie, Clyde, Dash, Love, and Morrison had all played in the eleven-a-side game the year before.

The New York rules were popular and on 30 July 1861 over 800 Hamiltonians watched the Maple Leafs, led by their catcher Bill Shuttleworth, defeat the Burlingtons on their grounds 33-27. The first significant inter-town baseball rivalry in Ontario began that year between Woodstock and Hamilton. The models for such rivalries were partially American but more significantly were based on those between local cricket teams which dated back to the 1830s between teams from Hamilton, Toronto, and Guelph. Inter-town matches in those early days were hindered by a primitive transportation system.

BASE BALL

ON THE FAIR GROUNDS

TUESDAY.

At 10 A. M.,

"Flyaways" of New York

vs.

"St. Lawrence," Kingston.

At 2 P. M.,

"Maple Leafs" of Guelph

vs.

"Live Oaks," Lynn, Mass.

The Programme further than this date will be announced hereafter.

Single Admission, - - -	25 Cts.
Grand Stand, - - - -	15 Cts.
Single Teams, - - -	10 Cts.
Double Teams, - - -	15 Cts.

Morning Despatch Print, Watertown, N. Y.

Baseball on the Fair Grounds in Watertown, New York, mid 1870s

There was a certain irony to these initial sporting encounters. At the same time as they gave territorial significance to towns only a few generations old by breeding an 'us against them' mentality, on a more profound level they were the glue that created a regional and in some cases a national sense of unity. By the 1860s the arrival of a visiting ball club by rail (it was a five-hour trip from Toronto to London) confirmed that Ontario towns were not only neighbours but linked in a common enterprise. In the mixed fortunes of regional competition was the implicit recognition that these new places made up of new immigrants or rural exiles were part of a larger whole known first as Upper Canada, then as Canada West, and eventually as Ontario. What baseball, unlike lacrosse and then hockey, was unable to do was create a national consciousness. That is, at least until the ultimate triumph of the Toronto Blue Jays.

On 3 September 1861 Jim Shuttleworth returned

to his home town along with the Woodstock Young Canadians, to play brother Bill's Maple Leafs. Woodstock's battery of Pascoe and Clyde weakened only in the fifth inning when they gave up 12 runs. McCann of Hamilton 'kept a sharp eye on the ball which he delivered with lightning speed,' but Woodstock won the two-hour, 45-minute match, 24–22, by virtue of three unanswered runs in the top of the ninth. A return match in Woodstock several weeks later resulted in an even more convincing Woodstock victory, 41–14. Hamilton rather feebly blamed the conditions of the field for their loss.

Woodstock's baseball supremacy

The best baseball in Ontario had now become a focus for heated celebration of local identity. Throughout the 1860s Woodstock would be the best Canadian team. Local amateurs and members of a pre-industrial working class, they included two coopers, two shoemakers, a carriage maker, a blacksmith, a tinsmith, a printer, a butcher, a carriage painter, and a grocer. McWhinnie, their first baseman, had been born in Ramsay, Ontario in 1834. He came to Woodstock at the age of 15 and in 1867 he would take over his father's newspaper. The centre fielder John Midgely was a Woodstock shoemaker who was born in England in 1843. Old occupations were disappearing and Midgely travelled west in the mid 1860s. He later worked on the railroad until a scandal involving 'evil doings in private cars' led to his dismissal. He went blind in 1908 and died in Chicago 14 years later. The experience of shoemakers Midgely and Shuttleworth is worth exploring for an explanation of how time was found to play baseball. Shoemakers formed a workplace group consisting of a laster, a heeler, a burnisher, and a finisher. If one worker was away the others could not perform their jobs and had to go home. Absenteeism due to family deaths was often quite marked, leading one wag to suggest that shoemakers had more aunts and uncles than any other profession. Even the death of a fellow workingman, which had once been the cause for a grand funeral procession, became instead an excuse for either attending or playing baseball. The ball team provided a chance for working people to establish communication networks within their occupational group. It was an integral part of the emerging labour movement.

The Woodstock team confirmed their dominant

position in Canada in 1863 with a convincing 29–6 victory over the Brother Jonathon club of Detroit. Jim Scarff, a local wagon-maker who had been born in England in 1839 and came to Woodstock in 1853, managed the Young Canadians' finances and travels. He had gone to California in 1858 with his father to search for gold but, finding none, returned to Woodstock in 1860 and from 1863 to 1870 was president of the baseball team. The team not only advertised the town to a regional market but enhanced Scarff's image among locals and he was eventually elected mayor four times.

The Shuttleworth brothers reunited in Hamilton in 1863 and, teamed with their fast ball pitcher McCann, challenged Woodstock. Scarff arrogantly told Hamilton's club secretary David Davies that Hamilton had no right to expect Woodstock to play them. 'After all, we're the champions,' Scarff said. Play they did, however, and in the series Woodstock won 41–25 in Hamilton and two weeks later won 26–12 at home. Jim Scarff and his brother Tom then took up a subscription and Woodstock fans excitedly contributed towards the manufacture of a silver ball to be awarded to the champions of Canada. The championship of Canada as such pertained only to the immediate vicinity of southwestern Ontario then known as Canada West.

Having proven themselves against all serious Canadian competition, the Woodstock team was curious to see how they might do against the best American players. In September 1864 word was received that the touring Brooklyn Atlantics would be playing near Rochester at the State Fair. Woodstock's challenge was accepted and the game was aggressively touted as the championship of the American continent. The introduction of the sleeping car into Canada by the Great Western Railway in 1862 ensured that Woodstock's journey would not, in the words of the day, be 'a chinese torture, in which victims die by being deprived of sleep'. Their competition represented the new breed of New York-based ballclubs. The Brooklyn Atlantics were a working-class club of 200 members with connections to the local Democratic Party. Its most influential backer was a local ward-heeler, A.R. Samuells, who owned the largest pool hall in Brooklyn with 50 tables and 500 seats. While the team claimed to be amateur, public funds were paid under the table to some of the better players. The Civil War failed to deter their operation. Their fans were rowdy and

ungentlemanly and included well-known gamblers and pickpockets.

Over 4,000 Rochester fans packed Jones Square Ball Grounds on September 22 for the Woodstock-Brooklyn game. They had hardly got settled before the Atlantics rattled the Canadians for five runs in the first inning and 21 in the second, including 12 before a man was retired. Fortunately for Woodstock balls that rolled to the fence were ruled to be singles. Brooklyn's 75–11 victory clearly indicated that Canadians had a long way to go before they could match the play of the New Yorkers.

Baseball's development in the United States was slowed by the Civil War, but there were no such restrictions in Canada. It was still three years before a formal country of Canada would become a reality, and organized baseball was largely reserved for Ontario. As early as 1861 the New York game of baseball was played between two picked nines of the Grenville club in Prescott—several hundred miles east of the Canadian hotbed of the game but still in Ontario. In 1864 the Dundas Mechanics following the example of an American organization for amateurs formed in 1857, helped establish a short-lived Canadian organization to promote and declare an annual champion to whom the silver ball would be presented. In Ingersoll some of the best bat and ball players dropped cricket and organized the Victorias club which lost by only two runs to Woodstock. Their first president and second baseman, Joseph Gibson, was a typical young ballplayer of the time. A man of limited education, he was Ingersoll's postmaster for 38 years. He was a staunch temperance supporter because of his father's drinking habits, and was an unsuccessful Conservative Party candidate in later years.

The end of the Civil War in the United States in 1865 did not immediately normalize affairs in North America. In fact relations between Canada and the United States were in turmoil. In 1857 Irish Americans had formed the Fenian movement to secure Irish independence from Britain. As many as 10,000 Civil War veterans had joined their cause by the end of 1865 and Canada was their closest British target. While they launched several sorties into Canada, the Fenians succeeded only in convincing reluctant provinces that they should join the new federation of Canada inaugurated 1 July 1867. That summer a major tournament in Detroit drew teams from that city as well as from Pittsburgh, Port Huron,

Woodstock, Hamilton and Ingersoll. Even though baseball, having finally supplanted cricket as the States' favourite bat and ball game in the early 1860s, was now termed America's national game, it was too well established in Ontario to be abandoned simply because of the American Fenian activity. Hamilton paid a $15 entry fee for the tourney's first division and a chance to win a $300 prize and a gold-mounted rosewood bat. Hamilton had borrowed two of Guelph's better players, Jim Nichols, a former cricketer and policeman, and tinsmith Bill Sunley, for the Detroit tournament and beat Port Huron 19-10 in the opener. McCann, Hamilton's pitcher, lost his touch in the next game and they lost to the Unknowns from Jackson, Michigan. Hamilton took home a runners-up gold-plated ball which became one of Bill Shuttleworth's prized possessions. 'The hard-fisted laboring men' of Ingersoll in their mid twenties entered the junior division made up of schoolboys and easily swept their four games. A gold-mounted rosewood bat and $100 was presented to their veteran captain Bill Hearn who along with the club's president Joseph Gibson had played the eleven-a-side Canadian game with the Rough and Ready club.

The Guelph Maple Leafs

The large cash prizes at such tournaments were part of the rapid commercialization of baseball. A fierce civic pride and win-at-all-costs attitude contributed to and was combined with a stream of verbal abuse about rival teams in local papers. Baseball had become the most obvious symbol of a competition between towns represented on the economic level by the confusing overlap of competing railroads. Guelph was keen to enter this competitive fray and they soon supplanted Woodstock as the dominant baseball town in southwestern Ontario. Guelph's local environment had offered few natural advantages for local entrepreneurs lured to the site by the promises and promotions of the Canada Company. A successful baseball team was one tool local businessmen would use to advertise their town's economy in the face of competition from rival cities like Hamilton and Toronto which benefited from a location on Lake Ontario. The New York rules of baseball had been introduced to Guelph in 1863 by Arthur Feast, a marble cutter and former member of the Hamilton Maple Leafs. The Guelph team was made up players

representing a variety of workplaces. They included four machinists, a labourer, a miller, a butcher, a policeman, a jeweller, a brewer, a tinsmith, and a Methodist clergyman. Jim Colson was the club's first president and shortstop. The team's pitcher Bill Sunley could 'talk baseball like any Guelphite,' one local columnist noted, 'till the sands of the desert grow cold.' Ten members of the team were Ontario-born. In 1864, their 'uniform' a red maple leaf on their work clothes, they defeated a Hamilton team whose line-up included many former Americans. *The New York Clipper* warned, 'The Canucks are not to be trifled with, and unless better teams are pitted against them in the future, the laurels may pass from the American boys to them.' The nucleus of Guelph's team was nurtured in the Union Baseball team, a junior club with such future stars as Charlie Maddocks, John Goldie, and Ephraim Stevenson. They would graduate to the Maple Leaf club of Guelph which was formally recognized with the passing of a set of bylaws in 1866.

A ferocious contest in Woodstock on the Civic Holiday of 4 August 1868 demonstrated the brutal seriousness of baseball to local esteem. Up to 500 people and a brass band accompanied the Guelph team to Woodstock. An early home run shattered a window in the nearby Agricultural Hall and this seemed to inflame the passions of the Woodstock fans whose team won the game 38–28. Local toughs roamed the stands picking fights and Guelph players were threatened. Guelph residents were furious and encouraged the club's new president William Bookless to mount a serious challenge for the championship. Meanwhile Ingersoll won a ten-inning game from Woodstock and claimed the Silver Ball. When the Scarff brothers hesitated, three weeks of accusations and threats of legal action followed. Woodstock finally relented but when they won the trophy back Ingersoll tried to pass off a silver-plated replica.

A year later at a three-day tournament in London, Guelph wrested the title and the $150 prize away from Woodstock. Several days of celebration followed in the 'Royal City'. The fans applauded the efforts of Jim 'Bunty' Hewer who had replaced Eph Stevenson. According to Guelph player Bill Sunley, 'Eph was studying for the ministry at the time and evidently his parents did not think playing baseball was a suitable preparation for that calling. At any rate just before the game his mother took him for a drive in the country.' Shortly thereafter, Guelph narrowly beat Dundas by scoring two runs in the tenth inning. With men on second and third, Stevenson protected the lead by sliding across the infield on his elbows to catch a pop fly, reminiscent of Billy Martin's diving infield catch of a Jackie Robinson pop fly in the 1952 World Series. A female fan said, 'Eph, I was so excited in that last inning, I could not watch the game, and so I turned and looked at the fence till it was over.' In jubilation Guelph's mayor Dr Herod kicked his plug hat like a football. The title was defended at London's Provincial Fair in late September, but Jim Scarff passed off a replica of the silver ball to Bookless and kept the real trophy for his private study.

In 1870, a 19-year-old jewellery apprentice, Bill Smith, joined his home town Guelph Maple Leafs. He started at third base but soon was one of the leading pitchers in Ontario. American teams were beginning to approach young Canadians about playing for them. Mike Brannock, born in Guelph in 1853, played with Chicago of the National Association in 1871. Other players however stayed in Guelph where they could pursue a career while sharing in gate receipts and tournament prizes. In contrast to Woodstock's fate in Rochester in 1864, the Maple Leafs' Canadian players were every bit the match for the Forest City Club of Cleveland who visited Guelph in July 1871. In two other games that season against Cap Anson's Rockford team, Guelph were beaten badly but recovered by September to defend their title against Dundas. In 1872, Guelph led by Bill Sunley's pitching beat Baltimore who at the time were third in the National Association.

One of the most amusing moments in Canadian baseball history occurred in the summer of 1873 when the champions of the National Association for the past two years, the Boston Red Stockings, visited Guelph. Four thousand men and women attended, some arriving by train from Brantford, Elora, and Fergus. 'One remark we wish to make,' noted the local base-ballist in the paper, 'is that the ball seemed a great deal more lively than the dead ball generally used.' Boston outslugged the Leafs 27–8. It was a game spoken of for years in the Royal City and one eventually

The Maple Leaf Base Ball Club of Guelph, Ontario, 1870. Canadian Illustrated News, *5 November 1870*

immortalized by the literary patron of the old west, Zane Grey, in his story, 'The Winning Ball' which appeared in the 1920 collection, *The Red Headed Outfield*. Grey described the adventures of a barnstorming ballclub forced to play an exhibition match in Guelph prior to a game in Toronto.

'We had never heard of Guelph,' Grey wrote. 'We did not care anything about rube baseball teams. Baseball was not play to us, it was the hardest kind of work and of all things an exhibition game was an abomination.' Grey's protagonists then described the Guelph crowd. 'Some 500 men and boys trotted curiously along with us, for all the world as if the bus were a circus parade cage filled with striped tigers.

What a rustic, motley crowd massed about, in and on that ballground. There must have been 10,000. The audience was strange to us. The Indians, half breeds, French Canadians, the huge hulking, bearded farmers or traders or trappers whatever they were, were new to our baseball experience.'

What draws the link between fiction and fact was the insertion, by the visitors, into the fictional game described by Grey, of a rabbit ball which had been a factor in the real Guelph-Boston encounter. In fiction however the offenders were not allowed to prosper and Guelph won. Grey shrouded the game in other ways but there can be little doubt that he got the rough details of the real game from his brother

The Maple Leaf Base Ball Club of Guelph, Ontario, winners of the 1874 World Semi-Professional championship in Watertown, New York where they defeated the Ku Klux Klan club of Oneida, New York in the finals. Team owner George Sleeman (centre) was a life-long baseball supporter. At 85 years of age in 1926 he was operated on for an 'internal complaint' and according to the Mail and Empire *'. . . disdaining the use of ether. . . conversed freely with Dr H.O. Howitt and the attending nurses ... reminiscing on the early days of baseball in Guelph and relating anecdotes in connection with the old Maple Leafs who were at one time amateur baseball champions of the world'—a reference to the Watertown tournament. Courtesy Guelph Civic Museum and the Sleeman Breweries (George Sleeman was the great-grandfather of John Sleeman who reestablished the family brewery in the 1980s)*

The Maple Leaf Base Ball Club of Guelph, Ontario, 1876—one of the earliest Canadian baseball photographs showing a baseball audience. Note the large number of female baseball spectators; courtesy Robert Stewart

Romer, a minor league player with Buffalo's Eastern League team, who passed through Ontario 24 years after the Boston-Guelph game.

Guelph were clearly the class of Ontario teams but they noted the hiring of two American pros by an Ottawa team in 1872. Harry Spence and Bill Jones joined Guelph in 1873. These were the first steps in the gradual subservience of baseball's Canadian development to American interests—a continentalist drift that would ironically receive a significant boost from George Sleeman, a man who proclaimed his loyalty to the Canadian amateur. Sleeman was born in St David's, Ontario in 1841 where his father ran a successful brewery and distillery. George played baseball throughout the 1860s and was cajoled into participating in old timers' games well into his forties. His first serious managerial role was with a factory team he formed at the family's Silver Creek Brewery in 1872. Finding his own employees to be inadequate he provided jobs in the brewery for some of the better local amateurs. Games were played on a field behind the brewery and the only shower was in a neighbouring pond. So successful was this venture that the Leafs invited him to assume their club's presidency in 1874. In so doing baseball in Ontario was marking the passage from an era in which players controlled their own destiny to one in which players played and a business class managed the team's affairs. Among those businessmen was Thomas Goldie who was involved at an early age as the team's secretary and whose brother John was one of the team's star players. Thomas, born in 1849, was educated at McGill and in New York and became manager of his father's milling company.

Boston returned to Guelph in 1874 and the July 1st holiday match put $562 into Guelph's bank account. Sleeman also organized Guelph's visit to Watertown, New York that year where they won what was optimistically billed as the World's Semi-Professional championship. Bill Smith emerged as the Leafs' new pitching star, replacing Bill Sunley. American professionals George Keerl and Hank Myers were added to the Leafs' roster and they had no trouble defending their Canadian title. Tom Scarff of Woodstock finally presented them with the real silver ball.

Guelph won again in 1875 but serious challenges were coming from both Kingston and London. In response representatives of Ontario's leading teams met at Toronto's Walker House Hotel on 7 April 1876 to form the Canadian Association of Ballplayers, with Sleeman as president. Guelph's competitors included the London Tecumsehs, the Toronto Clippers, the Hamilton Standards and the Kingston St Lawrence, each paying $10. They were required to play four games against each other that season. The National League was also organized that year in the United States and at least for the time being Canada had a baseball organization to rival the American one. In other sports Canada was able to retain an independent league structure. In baseball, however, the adoption of an American model in which a sharp line was being drawn between managers and players, and the importing of American professionals, guaranteed that it was only a matter of time before Canadian interests would be subservient to American ones. In the short term, however, baseball fans in Ontario were able to watch the greatest and most successful Canadian baseball team of the 19th century—the London Tecumsehs.

> There is nothing so galling to a Toronto baseball assemblage as a duck-egging at the hands of the Hamiltons.
>
> *Philadelphia Sporting Life,*
> *6 June 1888*

London's rise to baseball prominence

London was at the centre of the game's rise in Ontario. In the year of the Beachville game the British stationed a garrison in London on the site of present-day Victoria Park in an area bound by Clarence and Dufferin. It remained there, except for the period of the Crimean War between 1853 to 1856, until 1869, and members of the garrison played cricket. Despite its head start and its British leaning, cricket failed to hold its ascendancy though its early popularity may account for London's slow rise to the top ranks of baseball in Ontario. George Railton's city directory listed the London Baseball Club's formal status in 1855. Junior teams were playing by 1857. In 1868 the original Tecumseh club was formed from the old London and Forest City clubs. Its key organizers were John Brown, later City Treasurer; Ed Moore, part owner of the Tecumseh House hotel; R.M. Meredith, then only 21, but later a Chief Justice of the Supreme

Court of Ontario; and a well known local physician, Dr Morden. Other prominent members were Daniel Perrin, owner of a biscuit company, newspaper employees Richard and William Southam, and Harry Gorman. William Southam would later establish the publishing empire that bears his family name. Jim Jury, the team's captain and janitor at the high school, was one of the few working-class members. By the early 1870s commercialism had begun to replace the club's social function. President Brown was accused of betting against his team, a charge he vehemently denied. And by 1872 when the club was known as the Athletics, the minutes read that 'if a better player be found, he will take the place of the member of the first or second nine, who the committee may strike off.'

On 27 August 1874 the Earl of Dufferin dedicated the old garrison site for use as a public park, confirming its shared use by cricket and baseball players for the past twenty years. Over that period the bat and ball players had put up fences to keep out cattle which roamed city streets at will. Nearby residents however objected to the barricades across their short cut. In 1875 London returned to serious baseball competition with a team bearing the old name of the Tecumsehs. They imported their first American pro, George Latham, but still lost three games to Guelph who were derisively referred to as the Foreign Legion. A year later the rapidly improving and now semi-professional Tecumsehs played north of the old garrison site on the old fair grounds. London had imported Fred Goldsmith and were paying him $100 a month. Upwards of 7,500 fans watched the May 24th holiday game against the Maple Leafs. London's victory confirmed that the tide had turned in Ontario baseball. London and Guelph were so good however that the league's other teams couldn't compete. 'The question of the championship,' said the *Hamilton Times*, 'was a mere question of dollars and cents, and by this manner London has come to the fore. Baseball is meant to develop the muscle of youth, and not be a gambling speculation for roughs. Cricket has steered clear of this. Why shouldn't baseball?' Hamilton Standards were essentially an amateur outfit and their total season revenues hardly

International Association baseball match between the London Tecumsehs and the Guelph Maple Leafs in London's new park on 27 June 1877. Canadian Illustrated News, *14 July 1877 (artist C.J. Dyer)*

The Tecumseh Base Ball Club of London, Ontario, 1876: back row (L-R) Hornung (shortstop), Brown (centre field), Gillean (left field), Ledwith (third base); front (L-R) Dinnen (second base), Goldsmith (pitcher), Latham (captain and first base), Powers (catcher), Hunter (right field). Canadian Illustrated News, *15 July 1876*

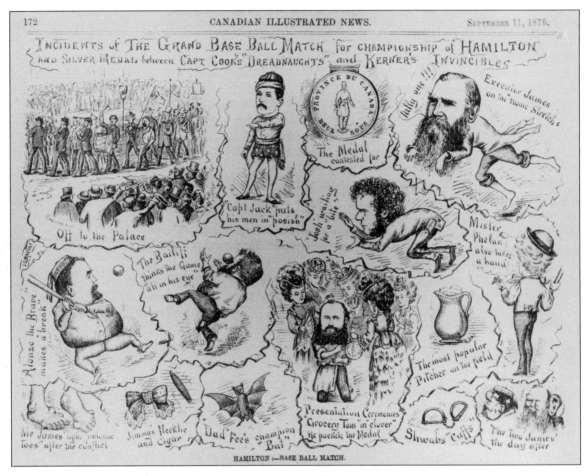

A local baseball championship in Hamilton, Ontario, 1875. Canadian Illustrated News, *11 September 1875*

reached $900. Spectators were mainly interested in high quality games played by professionals. Only 500 home-town Guelph fans watched the Leafs beat Hamilton but over 5,000 showed up for London's visit which the Tecumsehs won 10–7. Londoners jammed newspaper offices for telegraph reports, 600 took the train to Guelph for the game, and 400 packed the train station to welcome the team home.

Due largely to its successful commercialization, baseball had become the dominant game of southwestern Ontario. The game in Canada was in transition from its old style of play among members of the same club and the modern era of fierce rivalries. A scandal threatened the integrity of the league when it was reported that umpire Ed Moore, a Tecumseh director and umpire for an important match between London and Guelph, had bet a box of cigars on a London victory. Selection of umpires was still the responsibility of the home team. The Tecumsehs won

5–0 but it was later ruled that the game would not count in league standings. Yet the old times still persisted in some places. In mid season the Tecumsehs took part in a friendly match in Bowmanville, after which the London players and crowd sang musical favourites at a grand promenade concert and ice cream festival in the drill shed on Carlisle Avenue.

In mid season the Canadian Association discarded one of its last rule variations from the National League. Another part of the local control and uniqueness of the Canadian game disappeared. 'It has been found,' London general manager Harry Gorman said, 'that the rules regarding called balls and strikes are too favourable to the batter, and that Canadian games as a consequence, do not compare on the record with American games of the same class, played under American rules. Playing under different rules also leads to confusion and a change is very desirable.'

In mid-September, London closed out their

championship season with an 8–5 victory over Guelph. Each side charged the other with hiring professionals and the Toronto *Globe* editorialized that:

> should there be foundation to these charges, it would have a disastrous effect on the popularity of baseball. For as baseball loses the character of amateur amusement for players who love it for its own sake, and partakes of speculation in the engagement of mercenaries, as a game for gamblers, its sordid side is sure to extinguish its favour in the eyes of the Canadian public. Unless the game can free itself from the signs of dissolution, be prepared to see its sudden death as a Canadian pastime.

London and Guelph in the big leagues

Guelph's triumphant days as the leading team in Ontario had come to an end with the Maple Leafs' losses in 1876 to the London Tecumsehs. George Sleeman allowed Harry Gorman to play the leading role in negotiating the two communities' entry into the newly formed International Association. The Association represented a crucial threshold in the evolution of baseball management. The failed National Association which lasted from 1871 to 1875 had been unable to reconcile the interests of independent-minded team managers and players whose roles were never clearly delineated. Players felt little remorse in jumping from one team to another, sometimes in mid season. Club managers rejected a rigid schedule that would prevent them from pursuing lucrative exhibition matches. Both players and managers cared only for their own immediate franchise, and league governance was accordingly unstable.

The National League formed in 1876 deliberately limited ballplayers' authority, preferring to draw a strict line between labour and capital. Membership was limited to eight teams and a schedule was rigidly enforced. When Philadelphia and New York refused to play their remaining games after being eliminated from contention, the league kicked them out. It was a bold stroke but one that weakened the league by removing two of its biggest markets.

At the time there were probably fifty main-line teams in North America and no means existed to dif-

Tecumseh Club of London plays a visiting Chicago club in 1876. Canadian Illustrated News, *15 July 1876*

ferentiate between those we think of as major or minor league. Competing in a virtually open market for players were industrial centres of all sizes. Middle sized urban centres such as Lynn, Massachusetts and Syracuse, New York in the United States, and Guelph and London in Canada were engaged in a dramatic struggle for economic power. These were times of changing social patterns and labour upheaval as old professions disappeared. Civic life was dynamic. Opera houses were being built, trolley lines extended into emerging suburbs, and politically ambitious public figures such as Sleeman sponsored baseball teams. A ball club was a means to enhance an urban area's image and to attract new business.

In late November of 1876 L.C. Waite of St Louis invited all those clubs not aligned with the National League to contact him about a new association. Interested parties met in Pittsburgh in February. The idea of a league structure had little or no precedent in sports history. Entrepreneurs were wandering in unfamiliar territory. The Pittsburgh delegates rejected

the National League's exclusiveness and said 'that many true and tried friends had become so thoroughly disgusted with these irregularities'. They believed its restrictions on games between league and non-league teams was not only detrimental to the development of baseball but made no business sense. They might have concluded that the exit of New York and Philadelphia from the League proved their point. In any case they formed the International Association and adopted a liberal constitution which not only permitted members to pursue games with any other team but also welcomed teams that had no interest in competing in their championship.

The National League was a big city outfit that hoped to eliminate the existing network of independent baseball and by so doing eliminate the players' market and drive down the cost of salaries. In a few years they introduced the reserve clause which effectively protected a team's right to a player for life if he wanted to play in the league. They must also have recognized that smaller centres would never give up

1876 Canadian League game in Hamilton at the Crystal Palace Grounds. Canadian Illustrated News, *10 June 1876*

their love for baseball, which helps explain the evolving character of minor leagues beginning in the 1880s. In 1877 their only concession to outside teams was a carrot called the League Alliance. It would allow these teams to play exhibition matches against League members but they could not join the League. In return, under Clause 8, the League would have the authority to adjudicate all disputes between member teams. The Association trod carefully. It supported provisions calling for the respect of signed player contracts but firmly rejected Clause 8 arguing that it could handle its own internal disputes. The Alliance model however was the first tentative step in creating the structure of organized baseball under which teams are divided into major and minor league status.

The International Association spoke to a vision of major league sports which has much in common with the European soccer model in which countries with much smaller populations than the United States have supported a much higher number of competing major league franchises. The American college football and basketball structure is another indication that it is not so much a large number of franchises but travel and bloated schedules that water down competition. For its part the International Association despite the size of its cities was a major league. Ultimately, like the Union and Federal Leagues, it failed. In its willingness to open its membership and welcome teams from cities of all sizes, and most particularly in its international character, it was a far bolder and more interesting experiment than that of later failed major leagues like the Union and Federal which were simply following the National League's example.

Failure was not on George Sleeman's mind as he set out to recruit a team for the Association's inaugural 1877 season. His letter to Bill Craver (who was later banned from baseball for his role in a gambling scandal in Louisville) went astray, but he had better luck with Scott Hastings. Hastings' letter to Sleeman speaks volumes about the life of the itinerant ballplayer of that day. 'Now Mr Sleeman,' he writes,

for the, to me, disagreeable part of this letter. When you send me my money, start it about next Saturday or Monday . . . as I wish to make arrangements before I leave. And I wish you would make it a little more that it was the last time, if you will be so kind—I will have to leave some of it with my wife,

of course, and after I have paid my fare and other travelling expenses, I don't think that I will have quite enough to buy and stock a farm. I know that I am asking for more than was made an arrangement for, but sickness has cost me more than twice the difference I ask for, and you know that a person is powerless to prevent illness.

Sleeman was a typical International Association owner/executive. Of 129 officers of Association clubs there were 52 white collar workers in retail business or civic affairs including 12 present or future mayors, 13 blue collar workers, 11 players, 35 merchants ranging from shopkeepers to hotel owners, and 11 factory or mill owners. Sleeman was mayor of Guelph on more than one occasion. As well he promoted construction of the Guelph Opera House, owned the first street railway in town, and was at one time president of the Brewers and Malters Association of Ontario. A credit rating book of 1879 would estimate his wealth at between $40,000 to $75,000 but he couldn't compete with London's primary backer Jacob Englehart whose wealth was between $75,000 and $150,000. Englehart, born into a Jewish family in Cleveland, Ohio, got his business start as a representative of Solomon Sonneborn's notorious New York-based firm dealing in the illicit whiskey trade. Arriving in London in 1868, he sought to fit in by attending the Anglican Church. He eventually became head of a petroleum company with headquarters in that city and rose to the position of Vice-President of Imperial Oil Company. In 1905 he would become President of the Temiskaming and Northern Ontario Railway and the town of Englehart was named after him.

His Tecumsehs played on the old fair grounds site just north of the former military garrison field. Demands for use of the grounds during fair week created problems for the team and Englehart attempted to return to the garrison field which was now a public park. Local residents petitioned against the construction of a fence around the proposed ballpark. Englehart disputed their claim that a fence would be unsightly and said he was willing to spend $2000 beautifying the grounds. His business manager Harry Gorman argued that baseball provided enjoyment for thousands in western Ontario, many of whom arrived by train and spent their dollars in other parts of the city. An innocent, manly recreation was being threatened by an inveterate, unreasonable croaker, he concluded.

The croakers were several tavern keepers who wanted the team to re-locate to the Exhibition grounds near their establishments, where no doubt thirsty fans would gather before and after the game. By mid-April, with local politicians still debating the matter, the team leased a six-acre site in Kensington from the family of ballplayer Billy Reid. Kensington had opened as a subdivision in 1872 following construction of the Richmond Street Bridge and was a brief five minute walk from downtown and the railyards where the Great Western and Grand Trunk railways offered baseball fans in Port Stanley, Goderich, and elsewhere discounts for round trip tickets.

The grounds were dubbed Tecumseh Park. Though laid out on low-lying ground, they were fenced, sodded, levelled and equipped with forced water to give a permanent green look. Home plate was rather curiously placed in the northeast corner of the Park. Since games did not start until 3:30 in order to allow working people to attend, batters had to face the late afternoon sun. Broadbent and Overall managed the construction of stands and seats which consisted of a covered section for 600 and open bleachers. Mr Kitchen, formerly of the Great Western Railway, and at the time with the Montreal Telegraph Company, supervised the stringing of wire from the London office to the Park's press box, so that scores of games from around North America could be announced to fans. With the old courthouse visible in the distance along the third base line, the Park spoke to London's baseball past and its ever present respect for contemporary authority. The *Canadian Illustrated News* called Tecumseh Park without doubt the best for its purpose in the Dominion.

With two notable exceptions London's line-up was made up of itinerant American ballplayers. Managing the team, however, was Richard Southam whose brother William was to establish the publishing company that bears the family name. The Southams together with another newspaperman Harry Gorman and members of the Blackburn family who owned *The London Free Press* were early infected with baseball fever. Cricket may have had orthodoxy and class privilege on its side but baseball better reflected the spirit of the new world. It drew players and spectators and successful newspapers quickly followed the public trend. In London the game's roots were further enhanced by the fairly large colony of American expatriates who had made London their home during the American Civil War.

The other Canadian of some note with the Tecumsehs was Tom Gillean, a London jeweller. After a particularly good performance by Gillean in a preseason game against Hartford of the National League the *Free Press* noted, 'It pleased those who believe imports are not necessary, when good locals are available.' Gillean was strictly a second stringer on the team and his usual role was that of umpire in Tecumseh games. Such apparent conflict of interest was still tolerated, a quaint holdover from the days of gentlemanly intersquad games. Gillean would later umpire briefly in the National League.

Early season games in Canada have traditionally drawn small crowds due to cool weather and it was no different in 1877 as only 500 showed up to watch Association rival Pittsburgh when they visited Guelph and London. By the May 24th holiday however the baseball season was in full swing. In London upwards of 8,000 fans from all over southwestern Ontario jammed the new Tecumseh Park as the eventual National League champion Boston Red Stockings defeated London 7–6 in 10 innings. By June the *Canadian Illustrated News* was reporting that everyone gave up business for baseball when London played and defeated Guelph.

The cost of imported professionals was too much for Sleeman and in mid season he released his American ballplayers and finished the season with a contingent of local amateurs. 'You won't see our Club travelling all over the place throwing games,' Sleeman said, suggesting that some of his former players might not always have played honestly. Baseball was at a crucial turning point in its relationship with the paying public. At season end four Louisville players would be banished from the National League for throwing games, but the problem was becoming endemic throughout professional sports and threatened to erode the public's confidence unless something was done. Because ballplayers found it so easy to jump from one squad to another team owners had few recourses. The National League had tried one solution, that of making its games and league somewhat exclusive. In a few years they would find a second means by implementing a reserve clause that bound a player to one team for potentially his entire National League career. By such means transgressors could be effectively disciplined by the threat of a lifetime ban. It further served National League owners' interests by removing a free market and thus artificially helped lower salaries. It was a strategy that depended on the absence of strong competing leagues, a condition that did not always prevail in the 19th century. In 1877 London was arguably the match for any National League team and in August the Tecumsehs defeated the National League 1876 champion Chicago White Stockings (the direct ancestors of today's Chicago Cubs). Many Londoners saw the match as a world championship in that it paired their respective countries' national champions of the previous year. London beat Chicago 4–3 and the 1,500 London fans in attendance taunted Cap Anson with cries of 'out, out' when the White Stocking player objected to a close play on the basepaths.

In early October, Pittsburgh came to town for the climactic season finale, the winner of which would be declared International Association champion. A few days before, London's starting catcher Ed Somerville had died suddenly of pneumonia in Hamilton but accounts of the Pittsburgh game are sparse in referring to what one would have thought would have been a shattering tragedy. This may suggest that ballplayers and fans alike viewed the typical professional as a hired mercenary on whom little pity or care had to be expended. On a workday, 2,000 turned out to watch the Tecumsehs beat the Alleghenies 5–2 and win the first major league baseball championship ever recorded by a Canadian team. It would be 115 years before it would happen again and the victims, the Atlanta Braves, were ironically the descendants of the Boston team that defeated London on the May holiday of 1877. And of course it would be 116 years before the Toronto Blue Jays would match London's achievement of winning a major league championship on Canadian soil.

Over the following winter the National League extended a cautious membership invitation to the London Tecumsehs. In a letter to the management of the Buffalo club Harry Gorman stated, 'Our members seem inclined to accept the offer and I will go to Cleveland to see what they have to say.' The League was adamant in its strategy of limiting games with outside teams and Gorman balked at terms that would have meant an end to profitable exhibition games with Guelph and Buffalo. Besides, the International Association was taking on partners from larger cities and giving the appearance of having a more regulated and extensive schedule than that of the season before. The Tecumsehs walked away from the League offer. What difference would their membership have made? One can speculate

that a successful London National League team could have changed the entire course of professional sports in Canada, perhaps ringing an earlier death knell for lacrosse and even challenging hockey's later rise to prominence as the country's national game. We can further speculate on other Canadian cities entering major league baseball ranks and Canadian fans being swept up into pennant races and World Series matches generations before those of the 1990s. Even an unsuccessful London team might have provided an impetus for American major leagues to look at other Canadian cities as possible candidates for membership. Imagine for a minute however a London baseball team surviving and assuming a kind of relationship to the National League that Green Bay enjoys in the National Football League. It's a marvellously compelling thought but reality suggests that even had London joined the League its tenure would have been short. Other smaller places like Worcester, Massachusetts, Providence, Rhode Island, and Syracuse, New York did join the National League but faded from the scene long before the 20th century.

Supporting this conclusion was the tenuous nature of the Tecumsehs' 1878 season. By mid season despite the Association's apparent successes Jacob Englehart announced his desire to leave the team's Board to take care of his own mounting business pressures. The club's debts were a few thousand dollars and Richard Meredith intervened and raised $500 to pay players. Players had become disenchanted by the strict off-the-field standards set by their new manager Roscoe Barnes, a former Chicago White Stocking. After a victory over Rochester the players were accused of going on a drunken binge. In early July pitcher Fred Goldsmith pulled himself out a game complaining of a possible rib injury. The Tecumsehs blew a 4–0 lead and poor fielding allowed Syracuse to win in the ninth. The result looked suspicious to fans who were being served an almost daily dose of news about corrupt baseball doings in other places. When the Tecumsehs' local semi-professional amateur team, the Atlantics, signed Jim Devlin, one of the Lousiville four banished from the National League, confidence was further eroded.

Expenses were cut slightly when the team's eleventh man, Canadian Tommy Smith, was released. Goldsmith continued to complain about being overworked. 'It may be,' remarked the local *Advertiser* newspaper, 'that the public expected too much. The general impression is that their record was not because their opponents were too strong, but because the Tecumsehs did not throw a united strength into the game.' Harry Gorman tried to interest Hamilton investors in the idea of shared franchise. He was rejected and at a bitter shareholders' meeting delinquent shareholders were castigated.

Despite all this London were still near the top of the standings, but hopes for a pennant were fading. It was decided to pay off American professionals and replace them with local Canadian amateurs. The team limped to the season's end and then folded, ending a chapter in Canadian sports history.

CHAPTER FOUR

Maritime Baseball

Halifax, Nova Scotia and Saint John, New Brunswick, the largest cities in the Maritimes, have each promoted themselves as the dominant commercial and industrial centres of the region. In a contest between rivals of relatively equal strength it is a debate that is difficult to prove and ultimately rather pointless. Sports such as baseball however provide a means of making an abstract discussion somewhat more tangible. There is a winner and a loser. As often as not in the 19th century, the winning city was the one that opened its line-up to the best players regardless of their religion, social class, or professional standing.

Saint John's struggle to survive

Loyalist descendants in Saint John, New Brunswick were raised in a climate of bitterness fostered by losses sustained in their exile from the United States in the latter part of the 18th century. Their ranks consisted of civilians who often inherited the places of privilege in the future province, and a cadre of soldiers who had suffered most in the war and who lived in the lower cove of Saint John. Though often rivals, these two groups were both rabidly anti-American. Others however had had different experiences. In fact Saint John's first American links were positive, dating back to the arrival of several merchants from Boston in the 1760s. And the working-class Protestant Irish and Scots who arrived in the early 19th century to work at the docks and related ship-building and foundries had little reason to despise Americans.

Shrouded in Atlantic fog and ravaged by fire throughout the 19th century, Saint John, Canada's first incorporated city (1785), nevertheless flourished through much of that century. It benefited from the timber trade and ship-building. As well it functioned as a port of choice for Americans anxious to take advantage of the colony's preferential trading relationship with Britain. The underground economy developed by Americans smuggling goods into Saint John contributed to a boom and bust cycle of entrepreneurship. In 1851 Saint John trailed only Montreal and Quebec City in city size in Canada (it now ranks 26th).

A defining moment in the social life of Saint John was the great Irish famine of the 1840s which brought thousands of penniless Irish Catholics to the city's north end. The tensions of the old world found fresh opportunity in Saint John particularly since new emigrants were often willing to work for less money than the members of established communities. In 1849 a clash of Irish Protestant Orangemen and Catholic Irish resulted in 12 deaths. This tragic affair proved to be a kind of climax rather than a prelude to fates such as those witnessed in the late 20th century in Belfast, Northern Ireland. Animosity did not disappear but it found somewhat safer modes of expression. Baseball was one of these means.

Baseball's role in early Saint John

There are a number of accounts of early bat and ball games in Saint John including one that claimed ballplaying by leading citizens prior to 1840, and in

the period from 1853 to 1860 there is some indication of games resembling rounders and American
townball played by leading English and American
expatriates. Historian Brian Flood concludes that the
New York rules of baseball, which most closely
resemble those of today, did not reach Saint John until
1872. This is hardly surprising considering Saint
John's close connection to New England where townball and its later manifestation, the Massachusetts
game, was the dominant bat and ball game.

By the 1870s the game had become a popular
recreation among the city's working-class lads and in
1873 the Shamrock Association, a group of Irish
Catholic businessmen, established a ball club at York
Point. Their chief rival was the Saint John Club,
formed by the Protestant Saint John Athletic
Association.

A city-wide association formed in 1874 included
six teams. Many years later Mort Harrison of the
Royals team in that league recalled the team practices
in the city's South End. 'On a summer morning I
would go to the homes of the boys and rouse them
and we'd meet on the Barrack Green for practice
about 5 a.m. We had to go that early for we were due
to work at 7 o'clock and a day at the shop did not end
until late in the evening . . . '. By 1876 a Saint John
team, the Mutuals, were travelling by rail to Bangor,
Maine to compete in a centennial celebration match.

Ironically baseball's rising popularity corresponded to Saint John's gradually declining fortunes.
The great fire of 1877 destroyed 1,600 buildings,
only a few of which were protected by insurance.
Almost 1,300 were replaced within a year but the
commendable rebirth drained individual and corporate pockets of capital reserves that might have
allowed Saint John to finance the economic retooling
required by competition from other eastern seaboard
ports and the decline of the lumber trade.

In response to the fire and the city's waning economic fortunes, young men left the city to go west in
search of work. Some of these were the city's best
ballplayers and the game declined as a result.
Baseball survived these uneasy years but its bat and
ball rival, cricket, which had depended on the city's
aging élite for support, was virtually abandoned.

Between 1885 and 1890 fierce parochial battles
were waged on the ball diamond between the
Protestant Nationals (who became the Saint Johns in
1889) based in the eastern end of the city on
Rothesay Avenue, and the Catholic Shamrocks at

Shamrock Grounds on Lansdowne Avenue. In 1887
the city championship between the two was delayed
an hour because the imported Shamrocks pitcher
had yet to arrive from Maine. After much wrangling
he eventually played third base. The Protestants won
this game 4–1 before 1,200 spectators. At this time
teams were ostensibly amateur and stocked by local
players. A job was often found for better local players to ensure their loyalty and willingness to remain
in town. The Shamrocks' action changed that. Both
teams now began recruiting American ballplayers.

Beginning of inter-provincial play

Baseball was part of larger processes at work in North
American society. Foremost among these was an
attempt to create a larger public culture in which the
restrictive class differences of Europe would be eliminated. Baseball succeeded at least partially in forging
social consensus among all classes though this in no
way changed either the income of the fans or their
access to power. The common interest in the game
both by fans and players of all backgrounds was part
of the means by which some of the more odious
characteristics of Europe's social caste system were
blunted.

Equally important to the emerging public culture of North America was the way in which baseball
contributed to a civic and regional identity. As
Protestant teams played Catholic teams in Saint John
they began to see themselves as part of a civic unity.
In turn this helped forge a regional identity as was
witnessed in the formation of provincial leagues and
eventually in inter-provincial competition. By the late
1880s games were being arranged between the
Nationals of Saint John and the Atlantas of Halifax.
One series degenerated into name-calling as the
Nationals claimed that umpire William Pickering had
been paid off by local gamblers to decide the matter
in Halifax's favour.

Fierce competition in Saint John

In Saint John, the Shamrocks again lost the pennant
to the Nationals in 1888. They responded by hiring
a number of American professionals for the 1889 season including a former Saint John native, John
O'Brien, whose family had left the city following the
great fire of 1877, when he was 11. His baseball skills
had been developed in Lewiston, Maine. Yet again

the Shamrocks lost the pennant to the Saint John Athletic Club's team, now renamed the Saint Johns. Baseball interest had reached a fever pitch by this time but public confidence was eroded, first by gambling and then by the unsavoury reputations of imported players. Two of the most notorious were James Guthrie and Edward Kelly from Maine who brought with them a number of young women seeking work at a Saint John bordello run by Mattie Perry, known to the locals as 'French Mattie'. According to historian Colin Howell, Guthrie's travelling companion Lizzie Duffy had recruited an under-age Annie Tuttle. Tuttle's mother arrived from Bangor to retrieve her daughter. The police were called, the family reunited, and the infamous French Mattie left on the next train with Edward Kelly along with Duffy and Guthrie.

Baseball could not escape the fate of other Saint John commercial interests. As early as 1864 the *British Colonist* newspaper had noted, 'The Saint Johnian is eager, ardent, and untiring. He gives all his life up to business. He opens his shop or his office at an early hour, he risks more, speculates more, loses more, makes more; he fails in business oftener, but after failure he always manages to rise again and make another fortune.' Baseball was an extension of this environment that celebrated business acumen and risk within the community and advertised the city's features to other centres. Saint John citizens had known for a decade that their city was in trouble. In the first five months of 1880, 30 local businesses went under. Baseball enthusiasm peaked during this decade of decline. The attraction of apparently frivolous activity is often observed in cultures experiencing economic distress. Baseball also provided for a degree of continuity and community in people's lives, necessary at a time of dislocation when many long-time residents were moving either to New England or to western Canada in search of new opportunity. According to historian Brian Flood, 'On game day, thousands left their work to attend the contest. The Saint John streetcar company had to put on a number of extra

A women's baseball team, from Bocabec in Charlotte County, New Brunswick, c. 1895. Provincial Archives of New Brunswick, P71/27

A country baseball game at Perth, New Brunswick (note the spectators sitting on a Canadian Pacific train stopped on a siding).
Public Archives of Canada, PA 48831

cars to accommodate the ball crowds . . . housewives, doctors, clerks, lumbermen, merchants, and fishermen could be seen sitting in the grandstand or standing along the sidelines.' For fans who could not attend, a scoreboard was erected on the corner of King and Germain streets. The local baseball 'cranks' followed game results provided by the CPR telegraph. Information was swapped and bets were taken.

In 1890 a New Brunswick professional league had teams in Moncton and Fredericton along with the Shamrocks and Saint Johns. Professionalism and the intra-city rival of the two Saint John teams peaked that year. The Shamrocks won a number of early games. In response the Saint Johns hired E.C. Howe, a pitcher from Harvard University in Boston. The Shamrocks responded by hiring F.J. Sexton, a pitcher from Brown University, for $150 a month. On 21 August 1890 at the Shamrock grounds a crowd of mainly Catholic supporters taunted and tried to pick fights with players on the Saint Johns team. The atmosphere reminded many of the sectarian violence of forty years before. Three days after the game the Saint Johns withdrew from the league. Their Protestant business backers feared reprisals when the teams played at the Saint Johns' grounds. They were willing to support the game as long as it enhanced their business interests but they now realized that in times of economic distress the city could ill afford open religious warfare. Without their chief rival the Shamrocks saw little need to retain professional players. They were allowed to go elsewhere and, indicative of how good baseball had been for

that short period in Saint John, six of the former Shamrocks eventually played major league baseball.

There were giants in Saint John

Even if Saint John could not support high level professionalism, the sport was too engrained in the local culture to die. In the next decade baseball slowly rebuilt its base among young people and local amateurs and in the process successfully fought off challenges from lacrosse and cricket. Within the Catholic north end, youthful supporters of the Shamrocks formed their own team, the Roses, to play against a nearby seminary in 1893. In the Protestant south and east ends the Alerts were formed, but the sectarian edge to local rivalries was somewhat blunted by greater co-operation at the civic level. In 1897 the Roses' best player, Frank 'Tip' O'Neill and the Alerts' star southpaw pitcher Jim Whelly were united on a city all-star team that played Halifax.

Baseball in Saint John was evolving with its local society. At first it had represented fraternal organizations and close-knit rival business interests in the city. These individual ethnic units were not large enough to compete against either economic or baseball rivals in the region, and civic interests demanded that citizens unite regardless of their religious background or social standing.

Nevertheless the Alerts and Roses could not completely ignore or escape their past. The Alerts began importing players in 1899 and a year later both clubs were fully professional. The quality of play was superb.

Larry McLean, a future 13-year big leaguer who had family connections throughout the Maritimes joined the Roses. The Alerts signed Fred Mitchell, a future member of Boston's American League team. The Roses signed future Philadelphia Athletics star pitcher Andy Coakley in 1901. A year later the Alerts brought in Jack 'Cy' Coombs who in 1910 would win 30 games for the Philadelphia Athletics. In 1903 the Roses signed a former Cleveland star, Louis 'Chief' Sockalexis, but even he couldn't get a hit off Coombs.

Baseball in Saint John was beginning to repeat the conditions that had led to trouble a decade before. In one game in September 1901 pitcher Webber of the Alerts, upset by some razzing, roughed up the Roses' first baseman. In the confusion, fans poured onto the fields and fights broke out between Catholic and Protestant spectators. One of the players finally called the police. 'It was not nice for the people present, especially the ladies, and players should restrain themselves no matter how great the insult,' said a Saint John newspaper.

By mid season 1903 the Alerts' backers were losing money. Player salaries and ground rental were costing them $835 a month. The team folded and the Alert players as a group moved to Waterville, Maine where they defeated the best clubs in the state of Maine. Saint John's greatest baseball era was at an end though the amateur game survived. In 1911 the city was briefly represented in the professional New Brunswick-Maine League.

Halifax baseball

In its earliest days Halifax baseball was supported by many who claimed it had moral purpose. Temperance advocates played an informal bat and ball game in nearby Dartmouth in 1841 and the game surfaced again in written accounts in 1868 with the establishment of the Halifax Baseball Club at Doran's Hotel. Club president Dr A.C. Cogswell, writes historian Colin D. Howell, 'saw sport as a remedy to youthful idleness and indolence, and a force contributing to mental well-being and physical health'. Organized recreation, according to the *Acadian Register,* would rescue Halifax youth from 'gawking lazily at street corners to stare at passersby, lounging about drinking saloons, smoking and guzzling'. Proving that baseball could reinforce proper backgrounds, one of Halifax's most successful early teams, the Young Atlantas, represented the Young Men's Literary Association.

Baseball's public fate was subject to the world around it. In the emerging urban industrial society of the late 19th century, jobs and lifestyles changed almost overnight and survival fostered a rigid hierarchy of social roles. Many men held dehumanizing, one-dimensional jobs. Women raised children and their entry into the workforce was restricted. Those who were different because of colour or ethnic background were marginalized, often because of their willingness to work for less pay than established labourers. In general most sports conformed to this reality by retaining the distinctions between participants.

Baseball, however, was affected at a very early stage of its history by commercial interests and accordingly it was one of the first sports to exhibit features we associate with the 20th century. Of the 133 Halifax ballplayers who played between 1874 and 1888, clerks, labourers, and the unskilled made up 48.5 per cent of the total; 31.5 per cent were skilled workers either in flourishing or soon-to-disappear occupations such as those of cabinet maker, carpenter, tailor, blacksmith, machinist, gas fitter, printer, baker, plumber, cooper, and bricklayer. Students, merchants, and professionals who dominated other sports were only 22.7 per cent of baseball's participants.

Irish Catholics at the lower end of the social strata made up over 50 per cent of all players. Anglicans and black African Baptists followed. Local black teams, first organized in the 1880s, played throughout the Maritimes. There were barnstorming female squads like the Chicago Ladies who visited Halifax in 1891, beating a Halifax amateur team 18–15 before 3,000 paying fans at the Wanderers Grounds while other spectators looked on from the slopes of Citadel Hill. Leading feminist Grace Ritchie of Halifax argued that such tours supported attempts to develop recreation programs for women.

Indicative of working people's interest in baseball, says Colin Howell, was Thomas Lambert, a well-known Halifax labour leader and employee of Taylor's

> The public common is black every evening with men and boys playing ball.
>
> *Philadelphia Sporting Life,* 13 June 1888 on baseball in Halifax, Nova Scotia

Boot and Shoe Factory. Lambert came to Halifax in 1865 with the 2nd Battalion of the Leicestershire regiment. By 1869 he was an international officer of the Knights of St Crispin and in 1872 he became First Grand Trustee of the International Lodge. Lambert organized a factory ball team and was the team's scorekeeper. In 1877 he presented a silver bat and ball to the city champion Atlantas on behalf of the Mechanics of Halifax. Such working-class involvement duplicated that in Hamilton, Ontario.

Looking beyond Nova Scotia

As audiences flocked to local games revenue was generated to pay for better players. Demand grew for games with other Maritime centres and nearby New England towns to which many young Maritimers had emigrated in search of work. The completion of the Intercolonial Railway from the United States to Halifax in 1876 made it easier to organize these inter-town matches.

The two leading teams in Halifax throughout the 1880s were the Socials and Atlantas. They were members of the 1888 Halifax Amateur Baseball League and amalgamated a year later under the Socials banner. Jocko Flynn, a Chicago professional, was hired as their coach. The team visited Saint John, New Brunswick and were mocked in a city parade. One sign declared 'Little Sister Halifax. Haligonian Specialties. Fog in Summer, Harbour Skating in Winter.' Saint John won, getting its revenge for an earlier loss attributed to the bribing of the Halifax umpire William Pickering.

The Socials' season included 21 inter-city matches against semi-professional teams like the John P. Lovell Arms Company and the Woven Hose team, both of Boston. They won eleven games. A year later the visiting Saint John team deliberately lost a game in Halifax in order to draw a larger crowd for the return game in New Brunswick. The professional game was tarnished and it crumbled when the Socials eventually disbanded.

Amazingly the Halifax/Saint John rivalry was revitalized almost 10,000 miles from the Maritimes. In what may be the first record of baseball in South Africa, Canadian troops fighting for Great Britain in the Boer War regularly played baseball in their spare time. A Saint John, New Brunswick soldier W.L. Wright, with the 21st Troop of E Division, wrote home in 1901 describing a baseball match between those from Saint John and Halifax. The Fog Eaters of Saint John were led by Sergeant Globe, while Sergeant Hurby managed the Herring Catchers from Halifax. Players wore khaki shirts and pants with white caps and betting on the outcome was lively. Among the Saint John citizens playing baseball and serving in the Boer War was Art Finnamore who was

Young Men's Social Club, the Halifax champions of 1887. Players include Rob Davison, James Doyle, Howard Smith, William Smith, George Scriven, William Pickering, W.R. Scriven, Allan McDonald, Jack Graham, James Farquhar. Photograph Collection of the Public Archives of Nova Scotia

still coaching with the Black's Harbour team in New Brunswick after the Second World War.

Back home, Halifax baseball rebounded by the turn of the century with a new core of imports. Prominent among them were Larry McLean of the Halifax Resolutes and Bill Hallman, a former Philadelphia Athletic second baseman and a 14-year major league veteran. Hallman had played on the touring Volunteer Organist Baseball and Theatre Company team. Free agency was rampant as local teams raided each other. The local amateur teams suffered. The *Sydney Record* said that spectators preferred 'to watch a few experts whose business it is to play for the public amusement'. Players were promised 'a good salary, a lazy time, and the small boys' idol'.

The boom and bust saga of Maritime baseball continued. With the departure of these pros the quality of play deteriorated. Without the allure of the big

game in Halifax an opportunity existed for the game to grow in more isolated places like the coal-mining areas of the province. Places like Springhill became leaders in sponsoring baseball teams because they could offer good local amateurs a job in the mines and a chance to play ball. For working-class lads the choice was simple. If a variety of commercial interests ranging from industries to civic boosters were willing to support them, why say no. Between 1921 and 1945 the Springhill Fencebusters won eight Nova Scotia titles.

Baseball in the twenties

The Great War slowed baseball's development but the sport quickly revived. In Saint John by 1920 the old Catholic-Protestant rivalry had received new impetus. North end Catholics' team, St Peter's,

The Moncton Independent Baseball Club, 1922. Provincial Archives of New Brunswick, P75/22

included Ray Hansen who had tried out with Toronto's International League team the season before. Competition was provided by the Protestant Great War Veterans Association team.

Real interest was now focused on regional championships and St Peter's won the Maritime title from 1919 to 1923. With interest shifting to Maritime competition, the baseball spotlight in New Brunswick would focus on teams able to compete at that level. Within a few years that team would be Saint John's smaller neighbour, the community of St Stephen on the Maine border.

The pro game however had one major burst of organizational success between the wars and although this occurred in Nova Scotia it was far away from the game's early centre in Halifax.

Baseball and coal mining in Nova Scotia

In the entire history of organized baseball only five communities in Nova Scotia have had minor league teams in organized baseball and all of them were from Cape Breton.

Until the Strait of Canso was bridged by a 1,370-metre causeway in 1955, Cape Breton was truly an island. The highland scenery surrounding the inland Bras d'Or Lake had brought back memories of Scotland to its Scottish settlers and visitors from the time of Alexander Graham Bell. There was a harder life to be lived in Cape Breton, however, than that of landscape appreciation. It was found in the coal-mining and steel-producing region dangling on the island's eastern border with the Atlantic Ocean.

Baseball and coal mining went hand in hand in Nova Scotia. A local newspaperman, R.S. Theakston, had promoted the game in Pictou County in the 1890s, and it was popular in the coal towns of Joggins, Westville, and Springhill. It was a rallying point for workers' interests and a symbol of their community. By 1905 Cape Breton towns such as Sydney, Sydney Mines, Reserve Mines and Glace Bay were importing players to play in a colliery district league. Only the community of Dominion No. 1 (named after its local mine shaft and changed to Dominion by provincial statute in 1906) did not import players. The Dominion Coal Company hoped to discourage the type of enthusiasm that saw crowds of up to 800 attend local games. They complained that games starting before 5 o'clock disrupted work.

'Picnics have also contributed their share of adverse influence,' the Sydney Record noted, 'but baseball is the principal sinner.' Miners responded that the only other time to play was Sunday but they had religious objections to playing on the sabbath.

In 1906 sports promoter M.J. Dryden of Sydney proposed a Cape Breton professional league. As in other parts of the Maritimes the success of these professional squads was a double-edged sword. In the short term they raised interest in baseball. In the long run, when the professional game faded spectators were often unwilling to support local amateurs whose level of play was somewhat inferior.

Professional baseball comes to Cape Breton

There was prolonged labour unrest in Cape Breton throughout the early 1920s. British Empire Steel Corporation was formed in 1920 by combining Halifax shipyards and two competing companies, Dominion Steel Corporation and Nova Scotia Steel and Coal Company. The effects of monopoly control were felt on the labour scene and were exacerbated by the reliance of Cape Breton workers on company housing and company stores. Baseball survived and became an even more important symbol of worker solidarity. Grudgingly mine managers recognized the wisdom of co-operating with, rather than opposing, the public's interest. They supported an amateur league which lasted through 1935.

A year later an old pattern re-emerged as the league imported five American professionals. The Canadian Amateur Baseball Association refused to authorize its membership and the National Association of Professional Baseball Leagues which had overseen the minor leagues since 1902 was not prepared to accept it, so the Colliery League assumed an outlaw status for 1936. Among its better American professionals were 'Rube' Wilson, 'Shufflin' Bill Mitchell, 'Lil Abner' Tansey, 'Hustlin Irish' Clancy, 'Dizzy' Doyle, and 'Snooks' Mandeville. One of the few locals who remained was Tommy Jackson who had been born within the shadow of the Caledonia mine. He held mine manager's papers but the Glace Bay Miners baseball team obtained a leave of absence for him from the now supportive company, which often closed down when the team played.

The Cape Breton Colliery League entered the ranks of organized baseball a year later with a Class 'D' rating. Among its prominent leaders was popular local sportsman Starr MacLeod, president of the Sydney Baseball Club, and league president Juvenile Court Judge A.D. (Andrew Dominic) Campbell. The enterprising MacLeod who lived in Whitney Pier owned the Starr movie theatre and was also a Sydney alderman. His club's business manager Clyde Nunn was a future Liberal member of Parliament. They well represented the commercial underpinnings of successful baseball organization in Canada. Campbell was a more interesting study. His father, a great athlete, had arrived in Reserve Mines in 1895. A.D. followed in his footsteps coaching football and studying law at Dalhousie University. The law partnership of Harrington, Forbes and Campbell based in Glace Bay and Sydney was prominent in the labour disputes of the 1920s. Campbell represented the foremost union activist of the time, James Bryson McLachlan, a Scottish 'Clydesider' who had come to Cape Breton around the turn of the century. Campbell's partner Gordon Harrington had the support of miners as he and the Conservatives swept to victory in 1925 ending 43 years of Liberal rule. Harrington became provincial premier in 1930 and one of his patronage appointments was that of Campbell as a Juvenile Court Judge. Campbell wanted to keep young people busy and off the street, particularly in a decade of high unemployment. He worked closely with Roman Catholic organizations to sponsor parish-based sports. Baseball was the game of choice among the working people of Cape Breton and local team owners viewed Campbell as the perfect choice for their chief administrator.

In 1937 Glace Bay won the first pennant in the five-team league which also included Sydney, Sydney Mines, Dominion, and New Waterford. Eliott Atkinson who operated a grocery store in Glace Bay

recalled that the local team's park at the South Street Field was nothing more than a football field. 'There was no fencing, they just passed a hat around. Very few of the fans had any money unless they'd hoboed across the straits and got a job on the mainland. Glace Bay was largely a coal-mining town and conditions were quite poor. Sydney had the steel works and they were a little better off than the miners.' MacLeod's Steel City Club of Sydney playing out of Victoria Park defeated Sydney Mines whose home was the Brown Street Park in a two-game playoff and went up against Glace Bay. Both teams' line-ups were largely American though in game three Sydney's Murray Matheson held Glace Bay runless for five innings. After Matheson had left the game Tommy Jackson drove in the winning run with a triple. The two teams then struggled through four more games, one of which was a tie. In a climatic eighth game there was no score after seven innings, and 5,000 fans were on the edge of their seats. Each team scored in the eighth and then in the ninth Sydney loaded the bases with two out. Johnny DiRubio stroked a single to give Starr MacLeod's home team, the Steel Citians, the title. The team's player manager Guido Panciera was declared a creditable role model by Sydney management as he rigorously enforced a ten p.m. curfew. Matheson retired after the season and locals hoped that a young star of St Theresa's Intermediate Champions would replace him, but Danny Gallivan found ultimate fame on the other side of an announcer's microphone.

In 1938 Judge Campbell reported an even more successful season though Dominion withdrew in mid season. Glace Bay Miners, managed by former Brooklyn Dodger Del Bissonette, won both the pennant and the playoffs. The league however suffered from a failure to attract financial partnerships with

The Sydney Steel City baseball team, 1937 champions of the Cape Breton Colliery League

major league clubs. The St Louis Cardinals boasted working relationships with clubs in every Class 'D' League (there were 20 in the United States) except the Colliery. The highlight of the 1938 season was Merle 'Lefty' Settlemire's twelve-inning no-hitter for Sydney against the Sydney Mines Ramblers. Lefty's career major league record was 0–6 with the Boston Red Sox in 1928.

In the league's final season, weather conditions 'unparalleled in the history of Cape Breton Island', and the declaration of war in Europe, which cut short the fall season to allow Canadian soldiers use of ballparks for training, caused significant financial losses. The only all-Canadian league in organized baseball at the time had some noteworthy moments in 1939. Spalding's *Official Guide* credited Glace Bay and Sydney Mines with a world's record when they played four consecutive games all ending in ties, two going into extra innings. In all 38 innings were played between 22 and 30 August (the other dates being 24 and 26 August) and the score remained 12–12. This is five innings longer than the celebrated Pawtucket-Rochester game of 1981 which however had only one break. Sydney won both the pennant and the playoff and Judge Campbell was chosen by Judge Bramham, president of the National Association of Professional Baseball Leagues, to act as chairman of the commission which supervised the Little World Series between Rochester, winner of the International League, and Louisville of the American Association.

The league had stepped up to Class 'C' designation in 1939 and Douglas 'Scotty' Robb from New Jersey put together a more seasoned umpiring crew. Robb later went on to umpire in both the National League (1948–52) and the American League (1952–53). One of those he brought north was 28-year-old Chuck Whittle from Philadelphia. The umpires liked the league. It offered more money than leagues of higher classification and the tight geographic proximity of league teams lessened the travel burden and let them get to know the fans. Whittle recalled hearing about the German invasion of Poland on September 1 while returning from a ball game in New Waterford. He realized the league was doomed:

I remember meeting with Judge Campbell and we came to the conclusion that the Colliery League would never operate again with war upon us and American players sure to be restricted in their travel. I remember how sad it was seeing the players with

tears in their eyes when they realized they wouldn't be returning. They hated to leave Cape Breton. And even today when I run across players of the day, they always speak in favourable terms of their days here.

The Maritime Championship

While pro ball flourished in Cape Breton in the last years of the depression, baseball interest throughout the rest of Atlantic Canada focused on the Maritime senior baseball championship. This interest peaked in the thirties when the baseball team from St Stephen, New Brunswick were the 'New York Yankees' of the region. Allied with the border town of Milltown, Maine they won seven Maritime titles and nine consecutive New Brunswick senior titles. According to *Halifax Herald* sports columnist Ace Foley, 'there was more interest in those final series than in the World Series.'

St Stephen may have been the best senior team in Canada in the thirties but unfortunately there was no opportunity to test the claim. Originally formed in 1929 by a citizens' committee that included Whidden Ganong of the Ganong chocolate factory and future mayor Arlo Hayman, their exploits throughout the region were legendary. In the 1932 Maritime final against the Yarmouth Gateways second baseman Orville Mitchell pulled the hidden ball trick to retire the last Yarmouth player and the players then ran immediately for their cars as the Nova Scotia fans stormed the field in protest. In 1933 they endured a 13-day rain delay before outlasting the Springhill Fencebusters. A year later the team even played an exhibition game against the National League Boston Braves, losing 11–3 after surrendering nine runs in the first two innings.

St Stephen's reign lasted until the Second World War and raised baseball interest to levels it will probably never again approach. In the process the competition between teams from New Brunswick, Nova Scotia, and Prince Edward Island gradually broke down. Prince Edward Island was the first partner to finally admit in the mid thirties that they could no longer compete.

Baseball in Prince Edward Island

The history of baseball on the Island is not as old as that of Nova Scotia or New Brunswick. It dates back at least as far as 1889 when the Diamond Club of Charlottetown based at Victoria Park organized games with clubs in Summerside and Pictou, Nova Scotia. They went to Emerald for the 1 July Dominion Day celebration. Later that year the Lawn Tennis Club and Park Cricket Club played an informal six-inning game that was won, not surprisingly, by the bat and ball cricket specialists 25–5.

Baseball's development was slowed by the province's island condition which limited games with other Maritime centres. Lacking large cities, baseball in PEI found its support in small rural communities like Pisquid, Tracadie, Peakes Station (whose line-up included eight MacDonalds), Roseneath, Stanhope and Baldwin's Road. The following challenge was issued in the 18 July 1891 edition of the *Daily Examiner*:

> The Pisquid Baseball Club intend holding a picnic on their grounds near Pisquid Station on Thursday, July 23 on which day they are prepared to meet and play any other nine on Prince Edward Island. If a match can be arranged, it will take place at 2 o'clock on their grounds at Pisquid.

By 1894 St Dunstan's College had its own team. Opened in 1855 as a preparatory school for seminarians, St Dunstan's was one more example of the role baseball played in the Catholic Church's community development efforts throughout Canada. Also in 1894 the Charlottetown club played a Tuesday afternoon game at Victoria Park with a team from Pictou, Nova Scotia for the island championship. A year later Charlottetown narrowly defeated an island rival from Tracadie 11–10 and in 1896 the Charlottetown Baseball Club won the Island championship with a 26–17 victory over Pisquid.

The Abegweits

The sport's eventual success in the province however required some form of central organization. This was realized over a period of years as the Charlottetown Athletic Association, founded in 1891, re-organized as the Charlottetown Amateur Athletic Association in 1897 and built 'a facility for track and field, baseball, rugby, cycling, and other summer sports'. It eventually combined in 1899 with another club to become the Abegweits-Crescents Athletic Club. The incorporation was designed to cover the entire island and culminated a decade in which most competitive sports had been introduced into the province. The

Abegweit organization allowed Prince Edward Island to compete on more equal terms with larger centres in the Maritimes and occasionally in other parts of Canada.

The Abegweit Amateur Athletic Association was modelled on the Montreal Amateur Athletic Association formed in 1881 and the Halifax Wanderers Club, established a year later. This was a period of crisis in sports organization in North America. Amateurism was seen as a means of retaining the integrity of play against the onslaught of commercial interests which were often subject to bribery and corruption. For this reason individual sports were willing to sacrifice some of their independence for the protection of an all-encompassing amateur federation.

In sports like baseball, however, professionalism elevated the quality of play. Commercial interests had fought a successful battle to control the worst abuses of corruption but as the 1919 fixing of the World Series was to show the sport was not yet squeaky clean. Baseball was at best a reluctant partner in these amateur associations and generally went about its business with scant regard for any authority other than its own. With Prince Edward Island's limited population baseball supporters could not organize their own affairs, nor could they ignore an amateur association as important as the Abegweits. They gladly accepted its leadership. Baseball's amateur standing in the province may have diminished the quality of play but it did not affect interest in the sport.

The game's informal and spontaneous character was addressed in 1903 when the Abegweits leased space from the Charlottetown amateur association and built a ball diamond. Two years later the Abegweits had their own team and it brought to the game organizational expertise and spectator support already demonstrated in rugby, hockey, and track and field. Throughout that summer competition between the Abbies and a new team, the Victorias, aroused fan interest. A junior team, the Orlebars, with many future Abegweit players won the Charlottetown Business College Trophy in 1908 and by 1910 the Abbies and three other teams played a 15-game schedule in a Charlottetown senior league. Wearing their association's traditional red and black colours the Abegweits won the city's senior league title from 1911 to 1913. By 1912 there were three leagues in Charlottetown—senior, intermediate, and junior levels. The 1913 Abegweit team led by the

The Orlebars, junior champions from Charlottetown, Prince Edward Island, 1908; courtesy Charles Ballem

battery of pitcher Lou Campbell and catcher Ches Vaniderstine defeated the Connaughts, a new athletic association, which had replaced the YMCA team in the senior league.

PEI in Maritime competition

Baseball's golden age in Prince Edward Island was between the First and Second World Wars. This was the era of Lou Campbell, PEI's Athlete of the first Century of Canadian Confederation. Campbell played most of his baseball career with the Abegweits but in 1923 was induced to join the semi-professional Charlottetown Madisons of the Island Senior League. The team, a collection of stars, was sponsored by the Madison Manufacturing Company, makers of pool tables which could be seen locally at the Lambros Pool Hall. The Madisons, led by Campbell's pitching and his regular appearance in the field between pitching assignments, compiled a 23–2 record easily defeating the Summerside All-Stars and Campbell's old team, the Abbies. They won the Island title, trouncing Summerside 11–0.

Commercialism had raised the level of play and the Madisons boldly played exhibition games with out-of-province teams like the Springhill Fencebusters. The Madisons' glory was short-lived. Their sponsor questioned the monetary return to his business and they disbanded. In the process their success threatened the Abbies' future. Not only was professionalism challenging amateurism but individual amateur sports were seeking their own identity and governing bodies outside umbrella associations like the Abegweits.

For his part Campbell returned to the Abbies. Without a dominant team other centres were emboldened to join the provincial league and it soon included St Dunstan's, League of Cross, and the rural communities of New Glasgow and Toronto. Campbell's 4–0 shut-out against the Torontos on 17 September 1924 returned the league title to the Abbies.

In his final competitive season Campbell pitched the Rovers, based in the north end of Charlottetown, to the league championship defeating among others the Eastern Stars and the Anchors from Charlottetown's west side. While fierce rivals in local matters the three teams combined under the Abegweit umbrella to represent Charlottetown in provincial and regional competition. The 1925 island championship for a trophy donated by Lieutenant Governor Heartz featured a deciding game no-hitter by Vince 'Lefty' McQuaid as the Abbies beat Summerside 2–0. In the popular Maritime senior championship the Abbies endured a six week delay before losing the title in two games to a team from Saint John, New Brunswick.

A year later the Abbies lost a thrilling Maritime championship semi-final to Westville, Nova Scotia. Stores closed early and fans jammed the Abegweit Grounds to see their heroes lose the third and deciding match 10–8. The Abbies were provincial champions each year from 1924 to 1933 but were never able to win the Maritime senior title. In 1928 coached by Lou Campbell they won the Island title over Summerside. In mid summer that year they split a series with Amherst, Nova Scotia and won and tied their two games with Moncton in August. In the Maritime playdowns they were swept in three games by Springhill who eventually won the title defeating McAdam, New Brunswick 6–2 on 5 October. In 1929 'Putty' Connors led them to a two-game sweep of Moncton. In the Maritime finals they won the opener from the Yarmouth (Nova Scotia) Gateways. before losing three straight and the championship. In 1930 they lost in the Maritime semifinals to the eventual champions, St Agnes of Halifax. In an attempt to add to the regional rugby, hockey, and track and field titles won by the Abbies, the association scheduled a variety of exhibition games over the summer of 1931 and the Abbies led by Johnny 'Snags' Squarebriggs drew a bye into the Maritime finals. Unfortunately their opposition was St Stephen, New Brunswick. The Abbies whose season had ended a month earlier lost in three straight. In the Maritime semi-finals a year later the loss to St Stephen was repeated.

By 1934 the powers of dissolution were tearing apart all-encompassing sports associations like the Abegweits. With the formation of a Prince Edward Island Amateur Baseball Association the Abegweits' central role was no longer required. The new organization immediately dropped Island teams from the senior Maritime competition to the intermediate level and thus ended an era in which the Abbies had been challenged to raise their level of play an extra notch. The Abegweits' organization did not survive the 1930s, challenged as it was by the rise of independent sports authorities and to some extent declining levels of competition brought on by its retreat from senior level play in the case of baseball.

The Rovers Baseball Team, 1928 champions of Charlottetown, Prince Edward Island. Public Archives of Prince Edward Island, Acc. 2320, 83-6

Post-Second World War baseball in PEI

In the immediate post-war period Prince Edward Island's provincial government was challenged to play a greater role in providing athletic and recreational leadership and facilities. The Abegweit grounds were now a housing development and a ball diamond was built at Memorial Field in 1947. A revived Charlottetown city league had been formed in 1946 with three teams and they in turn formed an all-star team to play for the Maritime intermediate championship. Led by manager Fred 'Husky' McCabe they won the island title against Summerside, beat a New Brunswick squad in the Maritime semi-finals, and then took both games from the Sydney Mines Ramblers to win Prince Edward Island's first Maritime baseball title. According to historian Charles Ballem 'the late 1940s would be the strongest period for baseball in the history of the sport in the province.'

Re-organized as the Abegweits in 1947 the Charlottetown team were annually beaten by the Curran and Briggs team from Summerside who won the Maritime intermediate title from 1948 to 1950. Their 17 July 1949 12-inning 0–0 tie with the Abegweits is considered an all-time Island classic game. The Abegweits held their last annual meeting in the fall of 1954. The organizational torch was passed on to individual clubs and independent sporting organizations.

Post-war senior baseball in Nova Scotia

After the war the best senior baseball in the Maritimes returned to Nova Scotia where the Halifax and District League replaced the Halifax Defence League. The Halifax Service, Halifax Shipyards, Halifax Arrows, Truro Bearcats and Middleton Cardinals started the 1946 season. Cardinal players included Manny MacIntyre who in the winter would make up an all-black hockey line for the Quebec Aces, and Danny and Garneau Seaman. The Cards eventually lost a five-game series for the Nova Scotia senior title to the Truro Bearcats led by pitcher Philip 'Skit' Ferguson. There was no Maritime senior championship that year.

By the late forties and fifties Nova Scotia's best team was the Dartmouth Arrows whose short right-field fence earned them the title 'Little Brooklyn' in reference to their park's similarity to the Brooklyn Dodgers' Ebbets Field. They were followed by the Stellarton Albions who won three Halifax and District championships from 1951 through 1953. Maritime titles were often not held because seasons ended at different times in neighbouring provinces, a problem that had damaged Charlottetown's hopes in the twenties and thirties. In 1951 for instance the champions of the Southern New Brunswick League, Black's Harbour, disbanded at their season end so no regional championship was held.

In the mid fifties the Halifax league attracted a large number of American college players who could not play in the regular minor leagues. One of these was future major leaguer Moe Drabowsky who pitched for Truro in 1955. Stellarton's team was drawn largely from Wake Forest University in the United States. By the late 50s, however, the quality of senior baseball had begun to decline. Increasingly it was being challenged by softball as the dominant bat and ball game in many Nova Scotia communities. Softball's skill level allowed poorer players to play with good ones, it had no structure of organized professional baseball to compete against, and it required less field maintenance as fewer balls were hit into the outfield. In the first blush of post-war euphoria both games could find ample spectator support and player talent but as the fifties progressed young men sought out alternative amusements. The first victim was often baseball.

The Halifax and District League folded and with it the importing of American ballplayers—a familiar scenario in Maritime baseball. Local talent got a chance to play but the quality of play declined and with it spectator interest. By 1968 the Nova Scotia Central Baseball League too had folded and the game was in trouble in its former stronghold of Cape Breton. An independent team, the Dartmouth Indians, defeated Sydney Steel Kings, a junior team, in the Nova Scotia playdowns and then beat the Chatham, New Brunswick Ironmen in a three-game series for the Maritime title. A year later there was not even competition for the Nova Scotia senior baseball championship. In 1970 matters improved somewhat as Sydney's team made it to the Maritime championship where they lost to the Edmundston, New Brunswick, Republicans.

Baseball's revival in the 1970s was as much as anything fed by the nationwide interest in the Montreal Expos who came along at a time when the game's future in the country was in question. Senior leagues returned on both the mainland and in Cape Breton in 1971, and the Montreal Expos appointed former Nova Scotia baseball star Garneau Seaman to be their chief scout in the Atlantic provinces.

Baseball in the Maritimes will always be a boom and bust sport but its popularity seems entrenched with the recent success of several Maritimers like Rheal Cormier and Vince Horsman in the majors. Attempts have been made to revive the pro game in the Maritimes with the importing of former pros like Bill Lee, U.L. Washington, Pepe Frias, and Rick Wise, but locals with a sense of history know that such ventures will be short-lived though fascinating while they last.

Vern Handrahan

Maritimers' baseball allegiance, in keeping with the game's north-south axis, has traditionally been to the Boston Red Sox. Dr Bobby Lund followed the Sox as a boy in Charlottetown, Prince Edward Island where he played junior baseball in 1954 alongside a future major leaguer, Vern Handrahan. Years later Lund was a spectator in Boston's Fenway Park when Handrahan struck out the Red Sox slugger Tony Conigliaro. Dr Lund recalled that a fan behind him declared, 'Hey, that Handrahan guy is from Prince Edward Island.'

The island's only other major leaguer, Henry Oxley from Covehead, had an undistinguished four-game career in New York City in both the National League and American Association in 1884. We don't know much about him but it is probably safe to assume that his family left the island for the United States soon after his birth in 1858. There are few records of the game on the Island prior to the 1890s and Oxley had a better chance of developing baseball skills in nearby New England. Canada's smallest province had hosted the Charlottetown Conference of 1864 at which details of Canadian Confederation had first been discussed, but its small population made it a follower rather than a leader in sporting matters.

Handrahan's rise to the majors was improbable at best. Most children's baseball at least until the fifties was of the generally informal character found throughout most of Canada. Vern Handrahan was 14 before he played his first significant organized baseball with a Charlottetown midget team in 1953. The hard-throwing right-hander made the Charlottetown Juniors of the city league a year later along with Bobby Lund. Two coaches, Tom MacFarlane and Jimmy 'Fiddler' MacDonald, taught him how to stand

on the mound, concentrate on the hitter, hold the runner on first, and throw the curveball. Said Handrahan, 'We used to spend hours in Fiddler's back yard on Spring Street, and I learned the basics there.'

He posted a 13–2 record in 1956 and two years later he and Lund went to Stellarton to play with the Albions of the Halifax District League against teams loaded with American college stars on their way to the professional leagues. Here the Milwaukee Braves signed him with a modest $1000 signing bonus in 1958.

He played minor league ball in Wellsville, New York in 1959 where he roomed with Phil Niekro, the eventual winner of 318 big league games. Next year Handrahan played in Eau Claire, Wisconsin under a future Montreal Expo manager, Jim Fanning. By 1964 he was pitching for the Kansas City Athletics where he appeared in 18 games. He returned to the Athletics again in 1966 but most of his career was in minor league towns like Jacksonville, Vancouver and Rochester where his team won the International League title in 1964. His 1967 Birmingham, Alabama team included future Hall of Famers Reggie Jackson and Rollie Fingers. Handrahan was a regular starter, and Birmingham beat Duke Snider's Albuquerque Dodgers in the Little World Series, emblematic of minor league supremacy.

His salary never topped $7000 and in 1971 he was released by the Athletics. He returned to Charlottetown, married, and went to work for Canada Post. Handrahan coached junior baseball for several years and played in the Twilight League with friends from his junior days.

His was a most improbable rise in an era when most players were American-born and trained for major league competition from an early age. Coming from a small province with only a limited baseball record, he symbolized the unlikely origins of so many major leaguers, even if his stay was brief and is largely forgotten except in his home town.

Baseball in Newfoundland

Newfoundland's baseball history provides an ironic set of bookends to the story of baseball in Canada. Ironic because the province was the last to enter Confederation—in 1949, over a hundred years after the game had begun in the rest of the country—and because since its entry into Confederation it has not sent a player to the major leagues.

A copy of the birth certificate of Newfoundland's

only major leaguer, Jim McKeever, resides in the Baseball Hall of Fame in Cooperstown and lists St John's as his birthplace though subsequent research suggests that this may be a mistake and his actual birthplace may have been Saint John, New Brunswick. He had one big league season in 1884 with Boston's short-lived Union Association franchise and had a paltry .136 batting average in 66 at-bats. His parents were Irish immigrants and we might deduce by McKeever's death in Boston in 1897 that the family left Newfoundland for New England shortly after Jim's birth. Debuting in a major league line-up at the age of 22 he would have had to learn the game somewhere. In 19th century New-foundland the only bat and ball he could have been exposed to would have been cricket.

Newfoundland's ties were to the British Isles. Many settlers were from southwestern England where early bat and ball games similar to baseball were played. In all parts of Canada primitive bat and ball games based on rounders were played and we can be relatively certain the same applied in Newfoundland. In the rest of Canada these games laid the ground-work for later baseball enthusiasm and this growth was encouraged by nearby American examples.

In Newfoundland, however, the connections between the island and the North American main-land were at best tenuous. The itinerant maritime community which lived there had little time for games. These were the preserve of Newfoundland's ruling classes from Britain. A newspaper report in 1824 noted, 'On Tuesday last the members of the St John's Cricket Club entertained a large and res-pectable party at the Navy, Army, and Commercial Hotel to celebrate the anniversary of the institution. They sat down to an excellent dinner at half past five o'clock—after the cloth was removed toasts were drunk and the party broke up at a late hour.' By 1871 Newfoundland had its own team in the first Intercolonial Cricket Match in Halifax.

St John's historian Paul O'Neill provides a colourful picture of cricket around the turn of the century.

> By horse-drawn victoria, and on foot, crowds fol-lowed the road from King's Bridge, once called King's Pinch, past Ross's farm, formerly Jocelyn's, to the grounds. The journey ended at the wayside inn of Peter Rutledge's overlooking the pitch. . . . At Peter's, patrons could wash the dust of the road

from their throat with liquors, beer or milk. The two storey building was enclosed on all sides by a verandah where those with picnic baskets found shade. Inside, luncheons were served to the teams, the umpires, and scorers during an interval in the all day matches. The return to the ground of the elevens, led by the umpires, would be the signal for a ripple of applause from the spectators.

Cricket's 19th century popularity was built on a loose foundation. Newfoundland's physical isolation had made it difficult to organize the type of regional competition which appealed to spectators, and lack of internal communication until the completion of Sir Robert Reid's railway in 1896 limited the sport's province-wide growth. Generally unfavourable weather conditions limited cricket's season. And so, the game was not well enough established in Newfoundland to survive the retreat of English influ-ence. With the gradual departure of the garrison and the strengthening of ties with the mainland the stage was set for baseball's arrival.

These factors, particularly those of weather and communication, also explain why baseball was so slow in matching its growth in just about every other part of North America. O'Neill, however notes that the changing rhythms of the workplace were ulti-mately more favourable for baseball than cricket. 'A sportsman of the time [1911–12] observed that in the old days, when there was one mail a month, it was possible to play cricket, but with daily mail ser-vice in St John's, businessmen could no longer afford to take time off.' These realities of urban industrial life had condemned cricket as early as the 1860s in the rest of North America.

Soccer with its British ties and more acceptable hours of play was the first to replace cricket as the popular summer sport. Baseball, played sporadically in the 19th century, according to Frank Graham, the late Curator of Newfoundland's Sports Archives, was played informally as early as 29 May 1901 on the Parade Grounds at the corner of Merrymeeting Road and Parade Street. The game appears to have been introduced in a more organized fashion in 1904 by mainland priests at the now infamous Mount Cashel boys' orphanage. The Catholic Church was a strong supporter of baseball throughout Canada and priests were leaders in introducing the game to young boys.

At Grand Falls in 1910 the game was played by American and Canadian supervisors and technicians opening the newsprint mill. Managers played baseball in 1912 at the Dominion Steel and Coal Corporation's iron ore operation on Bell Island in Conception Bay. In St John's, baseball was played by workers who built and then managed the railway and communication business of the Reid-Newfoundland Company.

It was also popular among workers at the Imperial Tobacco Company; in 1913 G.G. Allen, vice-president of the company's Newfoundland division, presented a Silver Cup for the winner of the St John's Amateur Baseball League, the first such association on the island. League promoters included J.O. Hawvermale, the American manager of the Imperial Tobacco Company, employees of the Reid-Newfoundland Railway, and the Bank of Montreal. The Wanderers team, in blue and white uniforms, was made up of Canadians and Americans working in St John's. Rail workers of the Reid-Newfoundland Company in their red and white uniforms played for the Red Lions. Two Irish squads—the Benevolent Irish Society in green and gold, and the Shamrocks composed of West End fans and English expatriates wearing all-white uniforms, and likely divided on religious grounds—made up the league's third and fourth teams.

Games were played on Wednesday afternoons which, as in many small towns throughout Canada, was a legal commercial half-holiday. On this, the eve of the First World War, baseball fever had captured Newfoundland's sporting fancy. High winds and cold weather as late as 23 June failed to deter fans who paid a dime for admission. A local tailor, Mark Chaplin, donated a silk waistcoat to the first player homering in a league game. For a Red Lions game, Imperial Tobacco donated 50 Kismet cigarettes to any player who doubled, and 500 for a home run.

That summer the Reid-Newfoundland Company presented an inter-town trophy for a competition

The Shamrocks club of St John's, Newfoundland just after World War One. Newfoundland Sports Archives

All-star team from St John's, Newfoundland which played Grand Falls in 1920. John B. Orr, sometimes referred to as the father of baseball in Newfoundland, is sitting in the centre of the team. Resting his hands on Orr's shoulders is a former British Empire boxing champion, Mike Shallow. Newfoundland Sports Archives

between teams from Grand Falls, Bell Island, and St John's. The St John's team was essentially an all-star squad of the best players in the local league.

The war abruptly ended this initial baseball infatuation as the colony immediately fell in behind the British. Some teams managed to continue playing through this period but the game rebounded in the twenties at Shamrock Field. John B. Orr—connected with a Boston fish products company, and the man often dubbed the father of Newfoundland baseball— aggressively promoted the game, as did Paddy Grace who wrote a baseball column, 'Diamond Dust', for the *Evening Telegram*. But war again imposed its presence in 1939 and Shamrock Field was turned into a military camp. A new park, St Pat's Field, opened in St John's in 1947 in the valley below the northeast corner of Empire Avenue and Carpasian Road. It was equipped with bleachers and lights for night games, and here Corner Brook won the first McCormack Trophy in 1948. In 1969 the game was transferred to a modern facility at Wishingwell Park on Empire Avenue West, once the site of a city dump.

Isolation and a small population account for the failure of Newfoundland to produce a major league talent in this century, though Corner Brook's Frank Humber, a left-handed pitcher, was drafted out of Wake Forest University by the Los Angeles Dodgers in 1987. Despite a promising minor league beginning he did not make it to the majors.

Newfoundland claimed its piece of Canadian baseball history in 1993 when Rob Butler became the first Canadian to play for a Canadian-based World Series champion. Butler's connection is a little forced but one that islanders had no trouble celebrating on his triumphant visit following the 1993 Series in which he scored a run in game four and singled in game five. Butler's father Frank was from the village of Butlerville, a Conception Bay community of 200 families, so named because the Butler family had been the first to settle there many years before. His father had left Butlerville when he was 18 looking for work in Toronto. Here he married. Two sons, Rob and Rich, have both been drafted by the Toronto Blue Jays. Rob played on the 1988 Canadian Olympic team before working his way up through the Blue Jay system. Some may scoff at Butler's minimal role in the Jays' 1993 season but baseball is as much a matter of sentiment as statistics. The Butler family's story is a familiar one to Newfoundlanders and the world series victory by a native son's child may inspire a future generation of island ballplayers.

Postscript

Native baseball in Canada

It is uncertain when the native population of Canada was first introduced to baseball though one theory suggests that 19th-century reservation schools were an influence in parts of the United States. In Canada the Mi'kmaq people of the Atlantic provinces played a game called the 'old fashion' similar to rounders up to the time of the Second World War. It became a ritual of community solidarity. Baseball playing was taken up by more skilled athletes. Historian Colin Howell has found reference in 1877 to a game between Mi'kmaqs and a team of 'fat boys' in Halifax. By the turn of the century some American natives had entered the major leagues, the most notable being the Penobscot, Louis Sockalexis from Maine. A St Regis native, Joe Tarbell, from the Cornwall, Ontario area had signed with Cleveland's American League team in 1911 and played in the New Brunswick-Maine League in 1913. His ancestor Chief Tarbell had prevented many St Regis braves from joining Butler's Rangers during the Revolutionary War.

Native ball flourished between the two World Wars. In western Canada Jimmy Rattlesnake (1909–72) from the Hobbema Reserve in central Alberta made a name for himself as a barnstorming professional. Ironically that reserve was focus for stories by W.P. Kinsella, famed in the 1980s as author of a series of baseball novels, the most notable being *Shoeless Joe* which became the movie, *Field of Dreams*. Baseball

Canada annually awards a trophy named for Jimmy Rattlesnake to an outstanding player on Team Canada's senior team. Of Rattlesnake, author Brenda Zeman quotes Laurel Harney who said, 'Batters all felt Rattlesnake was dangerous; he made mortals out of heroes. But catchers used to feel like they was stickin' their hand in a sack fulla rattlers.'

According to Colin Howell baseball was being played on the Maritime reserves of Bear River, Eskasoni, Chapel Island, Shubenacadie, Milbrook, Membertou, and Big Cove, and Lennox Island in Prince Edward Island in the twenties and likely before. There were regional championships in the period; Big Cove won the Maritime Indian title of 1932 and in 1938 teams competed in Richibucto, New Brunswick. The white media couldn't resist the cheap headline of teams playing for the 'maritime scalp'.

In the thirties native teams such as that from Bear River in Nova Scotia's Annapolis Valley played outside the reserves against teams of Acadians in Meteghan, black teams in Yarmouth, and one of the Maritimes' leading amateur squads, the Yarmouth Gateways. Chapel Island played Acadians from River Bourgeois, Arichat, Petit de Grat and L'Ardoise. They once played the Reserve Mines club of the Cape

Ermineskin band member Jim Rattlesnake, who played amateur and professional baseball in western Canada and the United States. City of Edmonton Archives A87-151

Breton Colliery League in the late thirties. In 1951 the Lennox Island team won Prince Edward Island's Intermediate 'B' title, defeating the Peake's Island Bombers.

Howell notes that baseball was eclipsed after the war by softball; 'the post-war policy of centralizing native people on a few large reserves broke up the old teams and traditions. . . . By 1960 baseball on the reserves was a thing of the past.'

Less well documented but no less significant were those Canadian native women who played against white women's teams throughout the country. According to women's sports specialist Laura Robinson the most notable was a 1931 female softball team from the Six Nations Reserve representing Caledonia. Farther north and west, where softball had yet to penetrate, better native women played on reserve men's fastball teams.

In the post-war period, on the Six Nations Reserve near Brantford, Ontario softball was the game of choice for both native men and women. The Oshweken Mohawk Ladies Softball team was formed in the 1940s under manager Mel Squire Hill. Sunday games were delayed until after the young players had attended Olive Maracle's Sunday School. The team disbanded in the mid 1960s, to be replaced by a new generation. The 'new' Mohawks, led by their outstanding pitcher Bev Beaver, won many native and non-native championships, and in 1980 in Oklahoma they took the North American native women's title.

In more isolated settings, such as the Slave Lake area in northern Alberta, native baseball survived for a longer period. During the Second World War softball had been played against teams from the American army stationed in the region during construction of

1947 Oshweken Mohawk Ladies Softball team, formed after World War Two as native men and women on the Six Nations Reserve switched from the game of baseball. Top (L-R) Jean (Powless) Cooper, Myrna (Hill) Hess, Jean (Garlow) Martin, Marion (Hill) Green, Mel Squire Hill, Kay Squire Hill, Vivian (Smith) Miller, Dorothy Jamieson; bottom row Ruth (Van Every) Hill, Vera Martin, Dolly (Van Every) Hill. Photo courtesy of George Beaver

the Alaska Highway. Baseball rebounded when some of the better local baseball players returned from the armed services. One post-war team, the Slave Lake Indians, was largely native in composition; many of its players were descended from Métis driven out of Manitoba in the previous century. The team even had a female member, May Sinclair, who crossed the gender border to play hardball in the early 1950s. According to her brother, Sam Sinclair, 'She could not only make contact but hit the long ball. May chased balls in the outfield and usually caught them.'

Other teams in the Lesser Slave Lake League—the Drift Pile Reserve, High Prairie, Faust, Canusoe, and Canyon Creek—were made up of both natives and local whites. Sam Sinclair recalled that all-star teams of league players competed in some of the better tournaments in the Peace River area against teams from Edmonton. Prizes often exceeded $4000.

Sam's brother Walter, a southpaw pitcher, even played for the ostensibly all-black team from Amber Valley, Alberta. 'Being an Indian he was dark and fit in well,' Sam said. A few local players like natives Joe Courtroille and Dave Girioux played some minor league baseball in the United States. By the early seventies, however, baseball in the Slave Lake area declined as skill levels and competitive variety faded. In the 1990s the popularity of televised Blue Jays games revived interest in the game. Some local reserves like the Sawridge have become enormously wealthy through land settlements associated with the development of timber and oil resources. Local baseball supporters like Sam Sinclair hope that affluence will lead to a resurgence of baseball in northern Alberta.

CHAPTER FIVE

Western Canada

Part One
Alberta Baseball

In keeping with their sense that they were the centre of the Canadian universe, baseball organizers in Ontario throughout the 19th century would, without a second thought, use the term 'Canadian League' to describe their provincial associations. Consistent with such Canadian provincialism was the Western Canada League, formed in 1907. It consisted solely of Alberta teams from Medicine Hat, Edmonton, Calgary, and Lethbridge.

Located 215 km southeast of Calgary on the banks of the Oldman River, Lethbridge was one of the first places where the game was played in Alberta. It had been occupied by 500 generations of Blackfoot Indians prior to European settlement and the establishment in 1869 by Montana traders of the notorious Fort Whoop-up as a centre for trading whiskey. The North-West Mounted Police stopped the liquor business in 1874, but more changes were imminent when in the early 1880s an assistant Indian commissioner, Elliott Galt, discovered coal fields in the region. Elliott's father was Sir Alexander Galt, a leading colonizer and railroad entrepreneur in pre-Confederation eastern Canada. The elder Galt obtained the backing of London financiers and with the support of friends in Ottawa obtained the land grants necessary for incorporation of the North Western Coal and Navigation Company in

> He evidently regarded the exchange of the profession of baseball for the study of theology as a serious error in judgement, and in this opinion every inning of the game confirmed him.
>
> Ralph Connor, from *The Sky Pilot: A Tale of the Foothills*

Lethbridge. As appreciation to the fledgling Alberta community he donated a piece of land—the Square—in the central business section of the town. Here sports of all sorts were played in view of local residents shopping on the main street. The result was a community spirit unmatched in its ability to combine the everyday reality of work with play.

Baseball was brought to Lethbridge by settlers from the United States and Eastern Canada and was played at least as early as 1886. Undoubtedly the cosy pro-American sentiments of an Alexander Galt (who in 1849 had advocated annexation by the US) and the early influence of the Montana 'entrepreneurs' softened the sentiments of settlers for the adoption of what was by the 1880s the leading game of the United States. One of the first references to baseball in the *Lethbridge News* on 14 June in 1888 described a match with a team from Medicine Hat, a contest made possible by the development of a rail link between the two towns in 1885. The local paper had a flurry of baseball-related items in the last years of the 1880s. In late August of 1888 a fundraising campaign was announced to purchase a cup for a baseball competition in the

North-West Territories. A week later the local paper editorialized that a monetary prize would make more sense because it would let the winning team improve their grounds. A few weeks later, however, the unwieldy character of tournament play was chastised in a call for an organized baseball league.

In early March 1889 the Lethbridge club met to elect officers and the first match was played on the Square a month later. Weather bestowed both benefits and hazards on local baseball play. Chinook winds in late winter could eliminate ice rinks overnight and prematurely welcome the baseball season. Those same strong winds could disrupt summer sports, though baseball was less affected than games played with larger balls like soccer. Lethbridge's 1889 team was considered to be the finest in the Territories and only the lateness of the season prevented a match with a squad from Kamloops, British Columbia. As good as the team was, there were still some locals who had little understanding of the game and objected in the pages of the *Lethbridge News* to what they considered to be the unsightly nature of a catcher's fence constructed on the Square:

We have no desire to spoil the sport, but it is a good sound rule not to use your own privileges to the disadvantage or offense of your fellows, and we think the most enthusiastic baseball man will admit that the catcher's fence is a disfigurement to our square, and a hideous eyesore. Could not a net be arranged to answer the same purpose?

Bicycling, tennis, and even polo appear in an index of sports-related stories of the *Lethbridge News* over the 1890s but incredibly there is no mention of baseball again until 1900. This doesn't mean that the game wasn't played, just that it didn't have the prominence accorded by newspaper mention. Perhaps the most notable baseball event of the 1900 season was the appearance of the Boston Bloomers women's baseball team who enlivened many a prairie evening that summer. From this point on baseball reference was regular. Junior teams were being organized in 1901 and school baseball is mentioned in 1905. Scandalous reference to a game being influenced by gamblers appeared in 1906 and that same year indoor baseball was played at the YMCA gym. This game foreshadowed the later popularity of softball which did much to erode baseball's early advantage among participants and players. The Lethbridge Savages were organized in 1906 and by season end the papers were trumpeting a new 'Big League Ball

A Calgary team plays at Lethbridge in 1906 (the date is 27 January; the chinooks must have been blowing). Glenbow Archives, NA 1276-1

A ball game at the Alix, Alberta fair, 1910. Glenbow Archives, NA 205-13

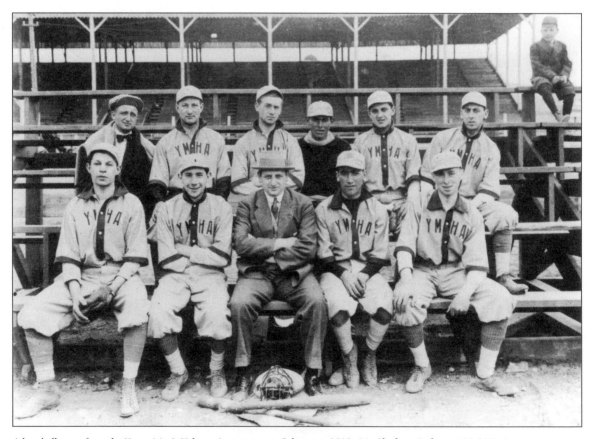

A baseball team from the Young Men's Hebrew Association in Calgary, c. 1912–14. Glenbow Archives, NA 2034-2

league' in western Canada. Local players got an early start on the season in December when the chinooks blessed Lethbridge with baseball weather.

Laying the groundwork for the Western Canada League

A successful league would require the participation of Alberta's leading urban centres but Edmonton, for one, had been somewhat slower than other parts of the province in taking up baseball. A survey of the *Edmonton Bulletin* for the late 1880s and early 90s reveals few baseball items and most were of the character of a 23 May 1892 game between married and single men at a local racetrack. An intercity match with Calgary was advertised in 1893 but subsequent papers provide no score.

There are several reasons for this apparent lack of interest. Following the merger of the Hudson's Bay Company and North West Company in 1821 Fort Edmonton had become the dominant centre for the western fur trade, but permanent settlement outside the fort did not begin until the 1870s. Even then American expatriates and Eastern Canadians who both brought the game to the region tended to settle in the southern parts of the province. The CPR went

through Calgary in 1883 but a transcontinental link for Edmonton in the form of the Canadian Northern Railway didn't reach the city until 1905 (the year of Alberta's recognition as a province). As capital of the new province and service centre for a huge agricultural region, Edmonton entered a boom period and by 1912 was home to over 40,000.

In 1907 the first significant provincial baseball organization was formed. Dubbed the Western Canada League, its membership as noted was entirely Albertan: Edmonton, Lethbridge, Medicine Hat, and Calgary. The Medicine Hat Hatters won the pennant. Egan, Lethbridge's shortstop, was fined $5 in mid season for punching an umpire but on a more positive note the team's star slugger Art O'Dea hit a league-leading .339. There was greater interest throughout Alberta in August when news arrived that the famous train robber Bill Miner had escaped from the New Westminster Penitentiary. Miner was a folk hero for his daring raids on the unpopular CPR—the line that, ironically, made it possible for baseball, the workingman's game, to be played between western Canadian cities.

This first Western Canada League lasted only one year but 'Deacon' White, founder and playing manager of the Edmonton team, would not be deterred. William Freemont White, a graduate of Northwestern

J. Lloyd Turner, left, and members of a Calgary baseball club. Turner came to Calgary in 1907 to play baseball. He became President of the Calgary Amateur Baseball league, and of the Calgary Athletics baseball team. Carl Zamloch is fourth from left; Ira Colwell, extreme right. Glenbow Archives, NA 460-14

Supporters' bus in Calgary. Sign reads, 'We are Canucks. Don't miss seeing the greatest baseball fight of modern times this evening 7:30 p.m. On fair grounds. We're going to Yank the American Grease Spots off the earth. No danger to non-participants.' Glenbow Archives, NA 1604-42

Calgary team of the Western Canada League, 1910. Glenbow Archives, NA 5329-17

University and a former schoolteacher, came to Edmonton as the member of an itinerant ball club that played a few games in old Fort Edmonton. White stayed behind and introduced the concept of twilight baseball: games starting in the late afternoon to allow working people to attend. Within a year White and another promoter, Frank Gray, were petitioning sympathetic baseball supporters in Calgary, Wetaskiwin, Medicine Hat, Moose Jaw, and Regina to re-form a new Western Canada League. They also approached Winnipeg and Brandon who until then had looked to the Dakotas and Minnesota for their competition. Deacon White was a persuasive salesman and eight teams—Edmonton, Calgary, Lethbridge, Winnipeg, Brandon, Moose Jaw, Regina, and Medicine Hat—signed on for the 1909 season in what could now truly be called the Western Canada League. Not only was it one of the earliest inter-provincial leagues in Canada but it spoke in concrete terms, as few other things at the time could, to the emerging sense of regional identity on the prairies.

Medicine Hat was again victorious. League meetings under President C.J. Eckstrom were taken up by spirited discussions on the loss of an important 1 July holiday date in Winnipeg when that town's local council told the baseball Maroons that Barnum's circus had first call on the ballpark. Considerable effort went into equalizing mileage with each team averaging 4,483 miles on the rail. Showing their affinity for the railroad, the Regina Bonepilers played at 'Railway Park', which had been expanded from 300 to 2,000 seats. All clubs lost between $1,000 and $3,500.

Decline and fall of the League

The league's reputation would be forever tarnished when in 1910 it banished one of its players for the sin of his apparent racial background. The year had got off to a good start in Saskatoon with the 30-strong 22nd Saskatchewan Lighthorse Regiment striking up the music for a 5:30 start. In the league's other Saskatchewan outpost, Regina, Dick Brookins from St Louis had been signed to play third base and the Regina *Leader* reported on 12 April that the 'Hard hitting Indian . . . will hold down the awkward corner.' Two weeks later he had three hits, two stolen bases, and two exceptional fielding plays in an exhibition game at the team's spring training site in Lacrosse, Wisconsin. Brookins started the season with Regina on 4 May but both Medicine Hat and Calgary protest-

ed his appearance on the grounds that 'he has negro blood in his veins'. Small numbers of blacks from Oklahoma were moving into Alberta at this time and the local populace was disturbed by this apparent threat to their racial hegemony in the province. Regina responded by confirming his native status. In the pecking order of baseball apartheid, aboriginal background was acceptable though hardly reputable.

Brookins continued to play and it was decided to refer the matter to baseball's National Commission which oversaw the operation of all teams which made up the structure of organized baseball. On 26 May the *Leader* reported that the editor of the *Sporting News* had written that on enquiries to the National Commission no dispute could be found to Brookins' eligibility; furthermore, 'Brookins' people are well known residents of St Louis and live in a neighbourhood exclusively for whites and are classed as white.'

Nevertheless league teams persisted in their complaints and on 2 June President Eckstrom following a protest of the Medicine Hat club declared that Brookins could no longer play despite the lack of a ruling from the National Commission. According to historian Owen Ricker the cause may lie in other events. In early June rumours circulated that Lethbridge's manager Chesty Cox was about to be fired, and Regina induced Cox to let Lethbridge's first baseman O'Hayer come to Regina. Benton Hatch, a Lethbridge official, angrily declared that he would ensure 'that nigger did not play any more'. Brookins heard the comment and would have gone after Hatch if he hadn't apologized. The next day however Medicine Hat protested Brookins' eligibility and Eckstrom banished Brookins. For his part Eckstrom had an interest in the Lethbridge club in that he owned the local park and let the League club use it free of charge. It hardly seems farfetched to suggest that Hatch spoke to Eckstrom and suggested the action that followed.

Medicine Hat had financial problems and dropped out of the league in mid season. Calgary was declared pennant winner of the second half of the season when Eckstrom was overruled after declaring Edmonton the winner. As for Brookins, he disappeared into the anonymous life of the itinerant ballplayer. A Regina paper declared 'neither Brookins or the club has received anything like a square deal, nor the faintest suspicion of British justice being in evidence the way the case has been handled.'

The league's moral authority had been sapped. A year later it was beset by deficit problems and in 1912

Saskatoon Quakers baseball club, c. 1914. Saskatchewan Archives Board, S-A 33

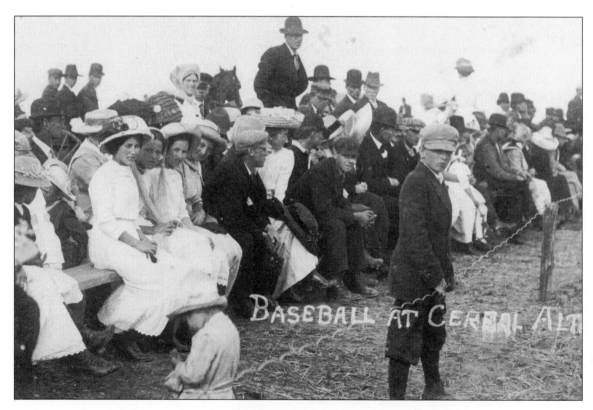

Spectators at a baseball game in Cereal, Alberta, 1913. Glenbow Archives, NA 2056-5

only four teams, all from Alberta, were in the league. In 1913 the playoff between two Saskatchewan teams from Moose Jaw and Saskatoon was disrupted by a nasty dispute between two Saskatoon players and an umpire. It was decided to replay the sixth game with Moose Jaw leading three games to two, but both teams objected and the season ended prematurely. In 1914 the league raised its individual salary maximum from $1500 to $1800 and teams were allowed to sign 14 rather than 12 players. The Saskatoon Quakers moved into a 6,000-seat ballpark funded by local entrepreneur J.F. Cairns. On opening day storekeepers closed their shops and the Quakers satisfied a sell-out crowd with a 6–4 victory over Regina. The Quakers won the pennant but the season and the league finished almost to the day that World War One began.

For citizens of western Canada and particularly Alberta, the league showed that though their settled history might be brief, they could match the civic attainment of eastern centres at least insofar as baseball was concerned.

Baseball and civic pride were both enhanced during that summer of 1914 by the visit of Sir Arthur Conan Doyle. Sherlock Holmes' creator often employed sporting items in his stories but on this speaking tour of Canada he demonstrated his athletic proficiency. Perhaps falling back on bat and ball skills learned at cricket, he tossed out the first ball at a baseball game in Jasper, Alberta between Jasper Park and Edson and then stepping up to the plate he 'hit the first ball pitched— for a homer, had he run', said *Saturday Night* magazine.

Deacon White's era

During the war Edmonton's Deacon White captained an army baseball team in Europe. One of his players was an Alberta farmer, Earl Thurston, an original member of the 49th Battalion. He was the signaller on duty when the ceasefire was transmitted on 11 November 1918.

White returned to Edmonton and by 1920 had a team in a revived Western Canada League. These were some of the most exciting years in that city's sports history. To open one season, World War One flying ace Wop May dropped a baseball on to Edmonton's Diamond Park from his JN4 airplane. In 1921 future

In 1919 World War One flying ace Wop May drops a baseball on Edmonton's Diamond Park from his JN4 airplane. Glenbow Archives, NC 6-4725

Former Edmonton baseball recruit Babe Herman and his Brooklyn Dodgers manager Wilbert Robinson

major league stars Babe Herman and Heinie Manush started their pro careers here, as did a future big league ump, Beans Reardon. Batboy for visiting teams was 13-year-old John Ducey, another easterner from Buffalo, New York, who came to Edmonton with his parents in 1910.

'I remember,' Ducey said, 'haunting old Diamond Park as a young fan in 1920. We kids would shag balls for the home and visiting teams during batting practice. Many good players besides Manush and Herman made it to the big leagues from the Western Canada League in those years—Tony Kaufman, a pitcher with Winnipeg, Oscar Mellilo, Winnipeg's second baseman, and Calgary's outfielder Walter "Cuckoo" Christenson.'

And from his home in Glendale, California Babe Herman had equally vivid memories: 'Edmonton in 1921 was a nice frontier town. Heinie Manush hit the first home run over the right field fence on opening day. They had movies of it and showed them that night at the theatre.'

Herman's frontier remarks weren't too far off the mark. To pay their way home at the end of the season, Herman and Manush played in an exhibition game against a Calgary team whose backers bet large amounts that their imported professional pitcher could handle the young Edmonton sluggers. The pitcher went head-hunting early in the game and an enraged Manush and Herman answered with seven extra base hits as Edmonton won easily.

The most fantastic frontier story concerns pitcher Vean Gregg from the State of Washington who had several outstanding major and minor league seasons between 1910 and 1918. He gave it all up to farm a quarter section of land he had acquired southwest of Edmonton in the Conjuring Creek district. He returned to the game after a few years, however, leaving the farm in the hands of a local man in the spring. Gregg retired from the professional game after the 1927 season but for many years he played for the local Conjuring Creek ball club. Other teams demanded that he not pitch. A loner, homesteading appealed to his private nature. But he left northern Alberta in the mid 1930s to return to the United States. He never came back.

Edmonton Eskimos baseball team, 1920. Glenbow Archives, ND 3-560

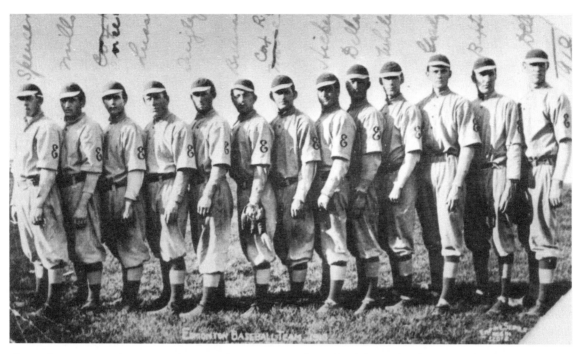

The 1910 Western Canada League's Edmonton Eskimos. Deacon White is fourth from right. The Provincial Archives of Alberta, A 7257

Deacon White remained a key figure in baseball throughout the 1920s and added to his sporting reputation by taking two Eskimo football teams to the Grey Cup. Like Gregg he also left the region during the Depression, spending his remaining years in Chicago.

Renfrew Park and Ducey Park

In 1933 a future National Hockey League President, Clarence Campbell, was instrumental in the construction of Renfrew Park very near the old site of Diamond Park in the valley of the North Saskatchewan River below McDougall Hill. (Much of the park was later reconstructed following a devastating fire in the early fifties.) American army teams played at Renfrew during the war. One Sunday afternoon game in 1945 between an air force team and a club from Fort Lewiston, Washington attracted 8,000 spectators.

After the war John Ducey was among those who organized the semi-professional Big Four League which flourished from 1947 to 1952. The league included the Edmonton Eskimos named for Brant Matthews' new hotel, Frank Wolfe's Edmonton Motor Cubs, and two Calgary entries: the Purity 99s sponsored by an oil company, and the Buffalos rep-

John Ducey, Edmonton's Mr Baseball, as an umpire in 1931. The Provincial Archives of Alberta, A 7276

resenting a local brewery. Among the league's grads was future Cincinnati Red Ted Tappe, as well as one of the few Alberta-born big leaguers, Glenn Gorbous, who created a sensation at a field day in Omaha, Nebraska in 1957 when he threw a baseball for a world record distance of 445 feet and three inches.

In 1953 Ducey was instrumental in returning the city to organized baseball with the admission of the Edmonton Eskimos to the Class 'A' Western International League (a league in which they had last been a member in 1922). With member clubs in Calgary, Vancouver, Victoria, as well as Idaho, Oregon, and Washington State, transportation costs were high and the league lasted only a few years.

Ducey continued his organizing efforts and in 1955 he entered Edmonton in a summer league for American college players. Among the players recruited from coach Rod Dadeaux's University of Southern California's team was the future Toronto Blue Jay General Manager Pat Gillick (then a left-handed pitcher), and the future Los Angeles Dodgers star (and an original Blue Jay) Ron Fairly.

Minor League ball returns to Edmonton

Minor League baseball returned to Edmonton in 1981 when Peter Pocklington, then at the peak of his local popularity for having brought the Edmonton Oilers and Wayne Gretzky to town, purchased the Ogden, Utah A's of the Triple A Pacific Coast League. They moved into 6,200-seat Renfrew Park, renamed John Ducey Park in March of 1984, six months after Ducey died, in recognition of his significant contribution to Edmonton baseball.

The team has provided Edmonton with many memorable moments. Future White Sox slugger Ron Kittle hit 50 home runs and 144 RBI for the 1982 Trappers. A year later the team beat the Reggie Jackson-led California Angels 5–3 in an exhibition encounter before 25,000 fans at a converted Commonwealth Stadium. Jackson cracked a loud double to a very deep centre field but future Blue Jay shortstop Dick Schofield responded with a home run for Edmonton. In 1984, Canadian Kirk McCaskill was the winning pitcher as Edmonton won the PCL title.

Since then Bert Blyleven, Fernando Valenzuela, and Steve Rogers have all tried to restore once great careers by pitching in Edmonton. Only Valenzuela briefly succeeded. And it was the demotion of Devon White to Edmonton by California during the 1990 season that led to his request for the trade which brought him to Toronto.

Edmonton's affiliation with the Angels ended after the 1992 season but not before local fans saw the minor league player of the year, Tim Salmon. In 1993 the Trappers commenced a two-year agreement

Kirk McCaskill from Kapuskasing, Ontario

with the teal-coloured Florida Marlins. Local fans were able to see big league prospects like Darrell Whitmore and Canadian Nigel Wilson.

Baseball in Calgary

Edmonton's Pacific Coast League rival in Alberta through the 1994 season was the Calgary Cannons. Somehow that seems appropriate. No Canadian community is imbued with so much American imagery as Calgary. That city's annual Stampede celebrates an era in which cowboys, trail herds, chuckwagons, and bronco busters roamed the open range. Yet the story is largely a myth and the Stampede, inaugurated in 1912, was a brilliant promotion by an American trick roper, Guy Weadick, who saw the possibility of transplanting the American rodeo to a Canadian setting.

The genuine Alberta cowboy of the late 19th century was in fact often an English gentleman, and hunting and polo were the popular sports in the foothills. Americans were a minority in the early days of settlement, and though there was a professional baseball team by 1889, the game was largely developed by migrating eastern Canadians.

Baseball's growth in Calgary owes no greater debt to Americans than that of other Canadian cities, but a visitor to Foothills Stadium cannot mistake the American atmosphere particularly during Stampede week when the town is full of out-of-towners in cowboy hats. The Calgary Cannons who play out of Foothills joined the Triple A Pacific Coast League in 1985 after Russ Parker paid $1 million for the Salt Lake City Gulls and received league approval to move them.

In early June 1993 the Cannons' major league parent, the Seattle Mariners team, was in town for a one-game exhibition. All attention was focused on manager Lou Piniella and Seattle's young super star Ken Griffey. His rapid rise to the majors a few years before had caused him to miss playing in Calgary for he went directly to Seattle from the Double A Eastern League. His big league aura was somewhat tarnished by an incident reported in the *Calgary Herald* the next day. A Calgary office worker asked Griffey to autograph a ball for Russ Parker, but the ballplayer told the messenger that the Cannons owner would have to ask himself. Griffey homered in his first at-bat and then left the game, retiring to a portable building beyond left field area which had little

resemblance to the palatial club houses of most major league teams.

Russ and Diane Parker's era

Russ and Diane Parker, Cannons owners, began their organized baseball affiliation in 1977 when they acquired a franchise in professional baseball's lowest rung, the Class 'A' Pioneer League.

Russ Parker was born in Moosomin, Saskatchewan in 1940 in an area known for its baseball. He played as a child before moving to Calgary in 1956 where he played junior baseball. As team manager he took a Calgary junior All-Star team to the national title in 1965. He then managed the Calgary Odeons of the newly formed Alberta Major Baseball League before becoming league commissioner in 1971, a position he held through 1976. In the early sixties young Parker couldn't afford university so he clerked at the Bay and became knowledgeable about office equipment and photocopiers. For a few years he sold equipment in Kansas City (where he was later to own an International Hockey League franchise) but homesickness brought him back to Calgary to distribute photocopiers, finally obtaining the Canon line in 1975. Calgary Copier Ltd, founded in 1969, eventually employed 150 Calgarians and became the fourth largest Canon dealership in Canada before the Parkers sold their interest in 1989.

The Parkers' Pioneer League franchise featured newly signed rookies and played at the municipally owned Foothills Stadium which had inferior lighting, bench seats, limited concessions, and a washroom beyond the outfield. A Player Development Contract was signed with the St Louis Cardinals and among the 1977 team was a future Toronto Blue Jay, Jim Gott. Two years later the Cards cut back on their farm system and so the Parkers obtained an affiliation with the Montreal Expos. It was perfect timing as the Expos were then entering their most successful years as a franchise, a period in which they would be known as Canada's team. The Expo connection produced division titles in 1981 and 1983 and among those who played in Calgary was Andres Galarraga in 1979 and '80. Generally, however, teams at the Pioneer level of baseball were located in much smaller towns. When Edmonton obtained a Triple A franchise in 1981 the Pacific Coast League hoped to find another Alberta team to help balance travel expenses. Calgary was a natural fit and a historic rival to Edmonton.

The Parkers had shared the Pioneer League investment but were sole proprietors in the purchase of the Salt Lake City team. The city invested $750,000 on upgrading Foothills Stadium and the team played its first Pacific Coast League game on 22 April 1985 after a three-day delay by a blizzard. A contest had been held to select a team name intended to 'reflect the character of the city'. The twenty finalists included several with obvious local reference such as Broncos, Buffaloes, Chinooks, Cowboys, Olympians (for the upcoming 1988 Winter Olympics), Pioneers, Stallions, and Stetsons. The winner betrayed in not too subtle a form the commercial reality of minor league ball. Announcing the name 'Cannons', the team called it a truly unique name, not having been used either in Calgary or elsewhere to the team's knowledge. The Cannons name, a press release said, was 'somewhat related to western and early Calgary heritage, as cannons were used by the Mounties during the development of the West'. 'Since we will be playing in a hitters' park, we looked for a name that would portray strength and power.' In fact the name best reflected the city's business character and in particular the Parkers' Canon copier franchise.

Few minor league franchises can survive without a major league affiliation to pay player salaries and the Cannons signed a two-year agreement to continue the previous relationship between Salt Lake City and the Seattle Mariners. The Mariners had checked out the California Angels' situation in Edmonton, found it favourable, and reported few concerns either with weather conditions or location. In announcing their affiliation with Calgary they confirmed their interest in putting a winning team in Foothills Stadium.

In keeping with their marketing success in the photocopier business, the Parkers aggressively pursued promotional opportunities. Broadcast rights were sold to Molson Breweries who were in a fierce national struggle with Labatt to control the Canadian sporting market. Outfield fence sign boards were sold at $4,000 each. 'We have no control over our on-field performance,' Parker said, 'because we do not own the ballplayers, and we have no control over the weather. So our plan was to market baseball, the fun of going to the ballpark, the fun atmosphere.'

The Parkers sensed a void in the family entertainment market and sought to exploit it through baseball with admission prices ranging from $3.50 to $5.50. Promotions included a White Christmas Day on 25 June, which featured Christmas carols, fake snow, and a visit from Santa Claus. In future years attractions were drawn from the barnstorming novelty entertainment sector such as the Dynamite Lady who crouched in a box which was then blown up. In the minor leagues the saying is never more true that no one ever went poor underestimating the public's bad taste.

Even with player salaries paid by the Mariners, the Parkers required an average of 3,000 fans for each of their 72 home games. By 1986 they were averaging 4,400 fans in a stadium seating 5,300, the highest percentage of seats filled over the season of any team in organized baseball; they sold out 13 times in their first season. In the process over $750,000 in merchandise ranging from bats to caps and toques was given away.

The Parkers began lobbying the city for stadium improvements and eventually a new park. After the 1987 season 1,300 seats were added behind the third base dugout and concessions and washrooms were upgraded. In the same off season Parker was recognized by Baseball Commissioner Peter Ueberroth for his role in making the Cannons one of the most successful Triple A franchises.

The great weakness of minor league baseball in the modern era is that it is often played at the behest of a major league team's particular interests. They often dictate a minor team's line-up either to showcase a player who may be traded or to give a promising rookie (such as Michael Jordan) a chance to play, even though better players may be available. Orders may also be sent down from the big club to let a pitcher throw a certain number of pitches regardless of game situation or its importance in the standing. Thus for fans of a minor league team the biggest attraction other than odd promotions and give-aways is not so much the final score as the chance to see future major leaguers. Among those who have played for Calgary are outfielder Danny Tartabull in 1985, a future American League batting champion Edgar Martinez in parts of 1985, 1987, 1988 and 1989, Omar Vizquel in 1988, 1989, 1990 and 1992, and Dave Fleming in 1991.

Old parks, new parks and the future of minor league ball in Alberta

The story of minor league franchises is often several years of popular support followed by declining interest either because the parent club fails to stock it with adequate talent or the local park loses its earlier charm. Such cannot be said for Edmonton's John Ducey Park. It remained into the 1994 season one of baseball's quaint hold-outs. Walking under the concrete stands with their great green doors and little pools of water and out onto the lush diamond, is like being cast back in time.

Writer W.P. Kinsella was so moved by his childhood experiences prowling the grounds in the fifties that he probably used Ducey Park as the model for his 'field of dreams' imagery in the novel *Shoeless Joe*. In the early 1990s, however, under the terms of a new agreement between the major leagues and the minors it was deemed unfit for organized baseball. In response Edmonton's City Council voted on 28 July 1992 to participate in the financing of a new 10,000-seat stadium with Trappers owner Peter Pocklington. This deal was confirmed two years later as part of an agreement to keep the Edmonton Oilers hockey team in town.

In Calgary baseball fans have retained their enthusiasm and in late July of 1994, the team, the city, and a federal National Infrastructure Program agreed to fund a major upgrade of Foothills Stadium to bring it to new Triple-A specifications.

Part Two
Saskatchewan

The film *A League of Their Own*, a fictional treatment of the All-American Girls Professional Baseball League, includes a player from Saskatchewan, thus recognizing the league's international cast. The real league lasted from 1943 through 1954 and while it was largely made up of American women, there was a sprinkling of Cubans. More significantly, just under 10 per cent (53 of 553) of the league's total roster was Canadian; of those, close to half (24) were from Saskatchewan (as well as 11 from Manitoba, 9 from Alberta, 6 from Ontario and 3 from British Columbia).

Baseball in Saskatchewan dates back to the earliest white settlement. In 1870 Rupert's Land had been purchased from the Hudson's Bay Company and transferred to the government of Canada. There followed a series of treaties with the native popula-tion who gradually retreated to designated reserves. Establishment of the North-West Mounted Police in 1873, a year after the first Dominion Lands Act, set the stage for significant settlement. According to Saskatchewan baseball historian David Shury the future province's first recorded baseball game was played outside Fort Battleford on 31 May 1879 between two makeup teams led by the Wyld brothers, Robert and Richard.

In this period homesteaders were attracted by almost free land. Saskatchewan's population totalled 19,000 by 1881 and 32,000 four years later, of whom half were British and 44 per cent were native. The arrival of the railroad in 1882–83 spurred immigration from the east. With them came baseball. There are accounts of games in 1881 between Colliston and Prince Albert. On 20 August 1882 an Ontario rail worker, Robert Martin, working on the CPR construction at Broadview, 275 miles west of Winnipeg, wrote in his diary that 'A crowd of fellows gathered to have an exciting game of baseball near the camp.'

Saskatchewan's future direction was settled in 1885 with the defeat of the Métis Rebellion and the eventual execution of their leader Louis Riel. The culture of the Métis, dependent on land and hunting, was eclipsed by that of British Canadians. One of their preeminent symbols was baseball. As historian Shury notes: 'From April 1885 until July 13, 1886 "E" company of General Middleton's troops stayed in the Qu'Appelle Valley while the rest of the troops were up north in the Riel Rebellion and "E" company helped pass the time by playing baseball.' Within a year baseball was being played casually in the Qu'Appelle Valley around Lumsden and the pitcher was the future province's first premier, Walter Scott.

Saskatchewan is a rural society of prairie farmland and grain elevators. Even its largest cities are like small towns. Its population had reached 900,000 by the 1930s and stayed there for almost a half century. When it entered Confederation in 1905 there were 23 communities strung along the national rail lines ranging in size from 500 to 6,500 residents. It was a society, says historian John Archer, in which 'pioneer wives were equal partners with their husbands in the work of pioneering.' Many women were former school teachers who married local homesteaders. They were leaders in pressing for adult education, libraries, and ultimately equality before the law and with that the vote. Nellie McClung was among the speakers at the first gathering of the Homemakers' Club in Regina in

Game in the Qu'Appelle Valley of Saskatchewan in 1886. Pitcher is Walter Scott, premier of the province from 1905 to 1916. Saskatchewan Archives Board, R-B 1859

Baseball team in La Fleche, Saskatchewan, c. 1920-21; back (L-R) C. Heas, Tim Pelissier, La Flamme, Emil Johnson, Leo Brunelle, E. Calfron; front, Frawley, A. Belcourt, W.E. Miller. Saskatchewan Archives Board, R-A 10,104

1911. As a child, McClung had railed against petty slights like the refusal of her Manitoba community to let girls play sports at a community picnic in 1882. The baseball game was restricted to married versus single men. Sports may seem an odd topic to raise in the context of human rights and the vote but small injustices are usually symbols of larger outrages.

On Wednesday, 25 July 1900 the Boston Bloomers, a team of seven women and two men (the catcher and the first baseman), came to Regina. Historically women had participated with men in the more informal bat and ball games of medieval society. (This at least partially explains why the English still think of baseball as the play of young girls.) The game of baseball as modernized in New York City in the 1840s corresponded to the sharp separation of gender roles characteristic of 19th century urban industrial society. Women were not encouraged to play and if they did, the thinking went, their play should be amongst themselves. And so as the game grew in popularity with the masses following the Civil War, women did take up the game. At an American women's college, Vassar, founded in 1865, there were two teams within its first year.

Rebelling against restrictions, working-class women formed barnstorming squads. In the mid 1880s the *Sporting Life* of Philadelphia decried the 'Disgraceful Conduct of the Girl Players in Georgia'. A Georgia report described a touring team of female teenagers from Philadelphia as 'jaunty in style, brazen in manner and peculiar in dress'. They were 'accompanied by some of the local swells', causing respectable ladies to get off the cars and await the next train. At big league parks, promoters offered free admission to women in hopes of leavening the largely male audience. It was a policy that wasn't always successful as attested by a *Sporting Life* description of the female baseball enthusiasts at an 1885 Polo Grounds game.

The rough and blackened finger tips of some of them show them to be working girls; others by unmistak-

able signs, even when they had not their children with them, showed that they were housewives and mothers; others still by their costly dresses and the carriages they came in, are seen to be well-to-do women and young girls, and a few are of that class of female gamblers and sporting women that has grown so considerable in New York City of late years.

The only option for professional female players was the open road. The Boston Bloomers were one such team. In Saskatchewan they defeated Moose Jaw's male team but their game in Regina was threatened by protest. This conservative community objected to women playing men but couldn't stop the game. A team of local males including members of the North-West Mounted Police defeated the women 13–12.

Baseball play among Saskatchewan women remained largely informal. Larger issues needed to be settled and these were significantly addressed near the end of the premiership of Walter Scott, with the granting of the vote to women in 1916. The former ballplayer warned that 'instead of any lightheartedness you will feel, it is time for serious consideration of these responsibilities.' In fact such a significant gain together with the privations of the First World War gave women the confidence to indulge in more frivolous activity.

At the same time the decline in the quality of women's baseball allowed for its greater democratization. Bloomer baseball had required skilled players but for most women there were few intermediary steps between the sandlots and a higher level of play. Women's baseball throughout North America declined between the wars and was surpassed by softball, then a simpler, easier game to play. A strong pitcher and a few skilled position players could dominate the game. Weaker players could improve their play at less demanding positions and experience no shame in striking out—most players did so.

In Saskatchewan, softball flourished by the 1930s. A generation of women were coming to age

> The seats from off the wagons were set around the place where the baseball game was played. The ball was a homemade yarn ball, and the bat a barrel stave, sharpened at one end, but it was a lovely game, and everyone got runs.
>
> Nellie McClung describes a prairie game in 1882; from *Clearing in the West*

Moose Jaw's Territorial Championship team of 1895 (note the integrated team's black player in the second row). Courtesy David Shury

Champions of the Saskatoon Ladies League, 1921. Saskatchewan Archives Board, R-B 4203

who would learn on the field and eventually participate in one of baseball's great experiments. One of these women was Christine Jewitt, a second generation softball player. Of the more than a hundred people including Jewitt inducted into Saskatchewan's Baseball Hall of Fame between 1985 and 1990 over 41 per cent, like Jewitt, were born outside the province (21 per cent from other parts of Canada, 14 per cent from the United States, and 6 per cent from Europe). Her birthplace was Leeds, England in 1926 but a year later the family emigrated to Canada, settling first in north-eastern Saskatchewan and eventually Regina where her ballplaying career began. As a 15-year-old she played Class A softball for the Regina Bombers in the Inter-City League.

St John's Presbyterian Church ladies' baseball team, 15 June 1921. Foote Collection, Manitoba Archives

Christine's great moment occurred in 1948 when she joined the Kenosha Comets of the All-American Girls Professional Baseball League. The league was the brainchild of Phil Wrigley, chewing-gum magnate and owner of the Chicago Cubs National League team. He had feared an erosion of fan interest in the game with the gradual exodus of ballplayers into the armed forces following America's entry into the Second World War. Wrigley's solution was a high powered women's softball league. At least its players would not be subject to the draft. Women's softball of this era had a slightly salacious edge. Its 'tough girls' reputation was spiced with rumours of either lesbian encounters or cigarette-smoking 'dames' with a man in every port. This image would be the height of radical chic in the 1990s but in its own time the enchantment of teams like Slapsie Maxie's Curvaceous Cuties and the Num Num Pretzel Girls of Miami Beach was lost on the general public. Wrigley aimed to clean up the women's game by combining good softball with decorum and the lessons of charm school. His scouts scoured the best semi-professional softball leagues in Chicago, California, Toronto and even Saskatchewan, and in 1943 the league's first games were played in four cities in the American Midwest. Lessons in makeup application and proper posture were combined with training in baseball fundamentals provided by former major league ballplayers.

In this hothouse atmosphere wonderful things happened. Under a regimen of daily play and practice, skill levels improved dramatically. Line-ups of talented female players were ill suited to softball's more limited game dominated by pitching staffs. There were few overmatched batters. Skilled fielders played at all positions. In response the league, with little apparent planning, began slowly moving towards the format of organized baseball. The evolution never reached a complete duplication of rules but came close. Between 1943 and 1954 the ball gradually shrank in size from 12 to 9¼ inches, the mound-to-plate distance went from 40 to 60 feet, the basepath distance from 65 to 85, and the pitching delivery changed from underhand to sidearm in 1946 and to overhand in 1948.

Christine Jewitt averaged .314 in her rookie season and hit Kenosha's first home run in their new Simmons Field ballpark. She lasted only two years however, returning to the Stewart Valley in Saskatchewan to marry and raise a family. They moved often but her softball career continued for another quarter of a century.

The loss of players like Jewitt dogged the AAGBL throughout its entire history. Like any league it relied on player continuity. Such a quality was found in Mary

'Bonnie' Baker, born in Regina in 1919. She was discouraged many times from playing ball. 'My father just hated that baseball, hated it,' she told writer Bob Boehm. 'I got whipped more than once because I stayed in [Wetmore] school to play ball.' She eventually joined the Moose Jaw Royals softball team. They toured the United States for six weeks in the late 1930s. She impressed American softball scouts, but turned down an offer to play semi-professional softball in Montreal when her husband objected. A few years later Hub Bishop who also worked as a hockey scout for the Chicago Black Hawks asked Baker if she'd like to try out for the new women's league. Her husband was now stationed in Europe. 'So when I got this offer from the All-American Girls' league,' Mary recalls, 'he didn't know I was gone till I wrote from South Bend.'

Baker was everything Wrigley could have wanted in a league player. She modelled on the side, represented the league on the popular television show, *What's My Line*, and was a rugged catcher who once almost came to blows with another Canadian, Gladys 'Terrie' Davis over a strike call. She even broke into the league's all male manager corps for one season with the Kalamazoo Lassies in 1950. She remained with the league until 1953 before returning home to Regina where she would become Canada's first female radio sports director and, later, manager of a curling rink. She, as much as any player, inspired the composite character that Geena Davis portrayed in the film *A League of Their Own*.

Terry Donahue from Melaval, Saskatchewan, born in 1925 was encouraged to play ball at an early age by her parents. They spent hours playing with Terry and her brother every day after daily farm chores were completed. She pitched for her school and played other positions for the Melaval Senior girls' team. William Passmore, owner of the Moose Jaw Royals softball team, was impressed by her pitching at the Lafleche sports day in 1941. The next year she joined the Royals and was a member of their 1942 Western Canadian championship team and 1943 provincial champion team. In 1945 she was selected as the southern Saskatchewan winner of a trip and try-out with the AAGBL. She won a spot with the Peoria Redwings. During her four-year stay the league made its almost complete transition from softball to baseball. Possibly for this reason she jumped to a Chicago-based professional softball league in 1950. She played there a few years before retiring and remaining in Chicago.

Arleene Johnson, born in 1924 in Ogewa, Saskatchewan, played girls' softball at both public and high school level in the area and sometimes subbed on

Mary 'Bonnie' Baker, a Saskatchewan sandlot graduate to the All-American Girls Professional Baseball League

Arleene "Johnnie" Johnson. ©1988 AAGBL Cards

the boys' team if it was short of players. In 1944 she moved to Regina and joined the Meadows Diamonds of the Inter-City Ladies' Softball League. She was scouted and signed to an AAGBL contract by Hub Bishop and joined the Fort Wayne Daisies in 1945. She was a superb third baseman who played in all league games in 1946 and 1947. Her career batting average of .164 was characteristic of a league never completely free of its roots in softball in which pitching traditionally dominates. Johnson returned to Regina in 1949, married, and following reinstatement as an amateur in 1950, had a thirty-year career as a softball player and coach. The *Calgary Herald* in 1951 said she was 'often rated as the greatest girl shortstop in Canada'.

Canadian women ballplayers: Margaret Callaghan, Helen Callaghan, Evelyn "Evie" Wawryshyn, Olive Little, Helen Nicol Fox, Dorothy Hunter. ©1988 AAGBL Cards

There were no Saskatchewan women left in the league when it played its last season in 1954. The Saskatchewan contingent, despite the fact that many married and raised families, did not retreat into unseen domesticity. There was no baseball to return to and so many played softball again. But it was now a softball whose quality and skill level had improved dramatically.

Postscript

Men's baseball in Saskatchewan had sent some of Canada's best ballplayers to the major leagues. The finest was Terry Puhl from Melville who left home at 16 to pursue what eventually became a 15-year big league career beginning in 1977. 'It's a different culture altogether going from Canada as a teenager to the United States to play baseball. Fortunately I had enough up-bringing to keep my senses and to keep on persisting. People don't realize that when I left home to play baseball that was basically one of the last times I did anything with my family in Canada.'

Puhl spent most of his career with the Houston Astros. During that time a former Canadian ballplayer, Tim Harkness, recalled Roger Craig telling him when they were teammates with the New York Mets that he wished more Canadians played the game because they could always be counted on when it got rough on the field. Said Harkness, 'Puhl is like that, he's made from good Saskatchewan stock, a guy who just puts his head down and goes to work.'

Ballplayers seem to spring from the Saskatchewan environment. Dave Pagan grew up on a farm near the small (population 100) community of Snowden, Saskatchewan. He didn't play baseball until the age of 14 when 'some kids got together and had a raffle to raise money for catcher's equipment and the senior team gave us some bats and balls.' In 1974 as a New York Yankee he played his best game beating Kansas City 9–2, but then he tore his rotator cuff. In 1976 he was part of one of baseball's most dramatic trades. Along with Rudy May, Tippy Martinez, Scott McGregor, and Rick Dempsey he was traded to the Baltimore Orioles. All of the new Orioles with the unfortunate exception of Pagan

> ## 'Sorry, Canuck.'
>
> — what Fergie Jenkins said to Terry Puhl after Fergie's pitch hit Puhl in a game at Wrigley Field

became stars. The Canadian retired to Nipawin, Saskatchewan where he worked as a carpenter.

Reggie Cleveland, the son of an RCAF corporal, grew up in Canadian military settings from Swift Current to Moose Jaw, Cold Lake, and Winnipeg. He pitched little league in Moose Jaw and was a javelin thrower in a Cold Lake high school. Though he was the second most winning Canadian pitcher in major league history and, in the 1990s, a pitching coach in the Toronto Blue Jays' system, he claimed that curler Ernie Richardson was his sporting hero.

Baseball's place in Saskatchewan is enhanced not only by its sandlot male and female graduates but its deep reservoir of organizations and supporters. One of its most notable is the Saskatchewan Baseball Association's Twilite Division formed in 1975 for players 35 years and older. Reputed to be the oldest type organization in Canada, its rules include the 'everybody bats' concept, unlimited substitution, pitching restrictions, and a tenth defensive player. The first twilite tournament was held in Watrous in 1976 and over the years two additional tournaments have been added attracting more than 40 teams.

Lastly, perhaps the most sincere attempt in all of Canada to preserve a regional baseball history is the Saskatchewan Baseball Hall of Fame based in North Battleford. David Shury has guided the hall since its first induction in 1985 and the baseball deeds of several hundred Saskatchewan residents are now stored in the hall's records. Shury's regular newsletters with tales of Saskatchewan baseball and notes on the passing of old-time heroes of local diamonds are a small Domesday Book of the game's heritage in the province.

Part Three
Baseball in Manitoba

'June weather flirted with Jack Frost' declared the *Winnipeg Free Press* as 700 shivering fans sat in the new stands of Osborne Stadium crunching roasted peanuts and preparing to watch some of the best baseball players in the world at the peak of their careers. In early June 1935 Winnipeg was visited by the great pitcher Satchel Paige and his equally renowned battery mate, catcher Quincy Trouppe. They were teammates on an integrated Bismarck, North Dakota team that played exhibition matches throughout the midwest. Bismarck's rivals that evening were the Kansas City Monarchs, one of the greatest black teams in baseball history. Their pitcher was the sensational Chet Brewer. It would be another decade before black athletes were allowed back into organized baseball from which they had been banned in the 1880s.

On this evening Paige struck out 17, while Brewer fanned 13. Both pitchers walked the bases loaded in order to get to a more favourable third out prospect. The crowd was not disappointed by the nine inning 0–0 result. In the next few days the four-game series would conclude with a Monarchs 2–1 victory over Bismarck and their white pitcher, followed by a doubleheader split on Saturday in which a rustier Paige outduelled Kansas City 11–4. The Monarchs won the second game 3–1.

Black barnstorming teams were baseball's refugees and for them western Canada offered the magnificent attraction of integrated rail transportation. People in the west were not necessarily more open-minded than those in other parts of North America but society there was still too new to have imposed as many restrictions. The freelance character of baseball in those days was attuned to this environment. It complemented the rhythms of a rural society.

Baseball's early days in Manitoba

The baseball affection of Manitoba citizens was hardly accidental. Frost-free days range from fewer than a hundred in some places to the four-month period in the Red River Valley. Manitobans, perhaps more so than southern Californians and those in the Caribbean, could identify with a game so associated with the return of spring.

Primitive bat and ball games had been played in the Red River Settlement but the modern game was introduced around the time Manitoba's provincial status was confirmed in 1870. The territory had been subject to the covetous interests of both Ontarians who had been moving into the region over the past ten years, and Americans such as the ex-Canadian George Shepperd who at the end of the Civil War argued that a little tightening of the economic ties

binding Red River and St Paul would cause the British outpost to drop 'like a ripe plum' into American hands. Land-hungry Ontarians of Anglo Protestant background however proved the greatest threat to the semi-indigenous Métis of mixed French Catholic and native blood. Unrest occurred and in 1870 troops under Colonel Wolseley were dispatched to bring law and order to the region. The soldiers generally took the side of the Ontario settlers and the emboldened farmers, whose ranks grew daily with the arrival of new settlers from the east, succeeded in driving the Métis into exile.

In one case a stone hurled at a fleeing Métis youth struck him on the head as he swam away, killing him. Such deadly aim may have been the result of ball-tossing skills learned in Ontario. Throwing an item with precision and technique is not an evolutionary talent but one largely developed in the artificial setting of a baseball diamond, of which there were many in Ontario by 1870.

Winnipeg's population swelled from perhaps a hundred in the late 1860s to 300 in 1871, to 2200 two years later, and 5000 by 1875. The first baseball was played here in 1870. Within four years the local baseball scene was lively. An organizational meeting in Winnipeg that year attracted 30 prospective ballplayers. During the summer of 1874 the Selkirk Cricket Club played the Garry Baseball Club and Garry was in turn challenged by the Moosehead club who offered to pay for their transport. There were two juvenile squads, the Mutuals and Pioneers, that season. Two years later baseball was part of Victoria Day celebrations in Portage La Prairie.

What was the source of this interest? Of 12,000 inhabitants in Manitoba in 1870 only 70 were from the United States. Like other features of Manitoba's first white settlement, baseball fever in the province had its origins in Ontario.

American influence soon appeared. By the end of the 1870s Winnipeg was the northern terminus of a railway out of St Paul, Minnesota that predated the CPR's arrival by several years. The CPR's eventual arrival brought not only new settlers but a summer rival, lacrosse. That sport's progress was slowed in 1885, according to historian Morris Mott, 'due to the fact that several prominent players were out of the city fighting in the North West uprising'. The troubles of 1870 between Métis and new settlers had returned for one last deadly time farther west. The rebel leader Louis Riel who had been in exile in the

United States, returned to lead a failed insurrection. His subsequent arrest and execution would inflame French-English passions nationwide.

Lacrosse had Canadian identity on its side. Despite its early history and growth in Canada, by the 1880s baseball was perceived as American. Nevertheless Winnipeg papers of 1886 were full of news on the National League and the American Association. One hotel provided daily telegraphic reports of games held around the continent. The telegraph provided an Ontario to Manitoba link as early as 1871, facilitating a local gambling industry and a somewhat widely held view that many sporting events had been tampered with by society's rougher edges.

Lacrosse's temporary decline in 1885 caused rail and hotel entrepreneurs to reassert baseball's hegemony. Baseball was a means to advertise their services, while players and followers of the game required transport and lodging. For those hotel keepers with access to telegraph services, baseball and an emerging gambling industry were a boon.

The CPR offered jobs to prospective players. Winnipeg's two leading clubs, the Metropolitans and the Hotelkeepers, realized that to compete they would have to import players. Completing this competitive structure was a team from Portage La Prairie. Portage had originally been a Church of England mission to the Indians of the Assiniboine and Qu'Appelle regions. Established in 1851, by the 1860s it was home to some of the most extreme ex-Ontarians. They had no time for Indian claims and actively sponsored additional settlement by Eastern Canadians. For a brief time in the early 1860s the 'Canada Firsters' of the area had established their own 'Republic of Manitoba' as a means of bypassing the Hudson's Bay Company's governance and usurping Métis interests.

Winnipeg Hotel Baseball Club, 1886, from the Manitoba Sun New Year's Illustrated. *Manitoba Archives*

Metropolitan Baseball Club of Winnipeg, 1886. Thomas Burns Collection, Manitoba Archives

Portage and three Winnipeg teams formed the Manitoba Baseball League in 1886, the first professional association of this sort on the prairies. Portage, dependent on local unpaid players, suffered a crushing defeat and dropped out of the league in early June. It was now a Winnipeg organization. Winnipegers welcomed the influx of professional talent. As early as 1878 the Prairie City Baseball Club had paid an American battery to play in an important Dominion Day match. These arrangements seldom lasted for long. Professionals of this era had developed an unsavoury connection with gambling and a somewhat libidinous character made possible by their generous salaries. On the other hand they offered a skilled and competitive brand of baseball.

The CPR discovered that its pockets weren't deep enough to compete and before the end of the 1886 season the railway team released their professionals

and dropped out of the league. The Hotels and the Metropolitans were another story. Buoyed by early season crowds sometimes reaching 2,000 of a population of 20,000, they scoured the lists of itinerant ballplayers for fresh talent. The Mets were run by the business and merchant élite of Winnipeg. At first they signed players from two local teams, the Winnipegs and the Ottawas. Soon they were paying an average of $250 a month plus room and board to eight professionals drawn largely from Ontario. At this time Winnipeg working people made only a few dollars per day. Two of the 1886 Mets, J.H. Barnfather and J. Brigham Young, had played the previous year with the Hamilton Primroses in the Canadian League. The Hotels club sponsored by several hotelkeepers imported five professionals from the American upper middle west. They were paid a total of $400 plus room and board.

Says Morris Mott, 'Winnipegers believed that only

a few centres in North America could boast of the calibre of baseball being played in their city. They took a good deal of pride in this because in the 1880s they still felt compelled to prove to Eastern Canadians and other North Americans, that their part of the world was "progressive" and civilized.'

High salaries and some players' questionable gambling habits soured many supporters by season end. In late summer Young, who would have one appearance for St Louis in the National League in 1892, taunted local fans. Another time he seemed to deliberately lose interest early in a game that his team eventually lost 24–2. His Metropolitans at 10–9–1 lost the pennant to the Hotel squad who finished 11–6–1. Manitoba's professional experiment did not survive beyond that season and amateur baseball returned for a 15-year period.

The amateur game

Paying players of questionable repute was costly but the practice would eventually find a new venue in the rural tournament. The tournament system was better suited to rural society than the urban leagues because schedules could be adjusted to local conditions and payments reserved only for the winning team and runners-up. Throughout Manitoba and Saskatchewan grain elevators and small service centres were spaced at 30-mile intervals along the CPR. From the late 19th century through the first decade of the new one farmers and their rural society would prosper on some of the most fertile land in western Canada, blessed as it had been by the retreating glaciers. The rural communities' economic ascendancy was in gradual decline by the end of the First World War but

in Manitoba farmers and town dwellers outnumbered city dwellers until 1941. This was a period in which a baseball diamond and a local team were often the first visible symbols of a vibrant community life. The quality of the games and their results are largely the stuff of

Anonymous Manitoba baseball player, 1914. Foote Collection, Manitoba Archives

memory and rural myth but these roadside attractions loaded with so much local meaning created a powerful community spirit. Unlike one-industry towns where baseball was sponsored as a means of controlling the enthusiasm and spare time of workers, on the prairies it became an extension of the largely independent lifestyle of farmers. It thus had a more authentic and hence larger life.

The Queen Valley White Stars formed in April 1908 were esteemed in the local community not only for baseball but for bending the elbow. Team records describe a walk to the Brokenhead River, two miles east of town, as a good place for 'glass blowers' to get back on their feet after too much drinking. One inebriated player was bitten on the leg by a bear chained to the front of a local hotel.

Before the First World War in the village of Waskada just above the North Dakota border in the southwest corner of the province a local bank manager sponsored a semi-professional team. Banks were either a farmer's friend or greatest enemy. In this case the civic-minded official had apparently pocketed some team expenses, and the ballclub soon followed his quick departure.

Four small centres northwest of Brandon—Hamiota, Oak River, Crandall, and Miniota—had their own league. Baseball had replaced lacrosse as the dominant recreation of the area. Each team deposited a $5 entry fee and a league rule declared that a player had to reside in the community for at least 10 days prior to a game. Admission to a game was 25 and 15 cents: women were allowed in for free. The First World War ended baseball in many small towns but in Hamiota a three-team intra-village league had squads named after the major league Tigers, Cubs, and Giants. Tournaments gave the town the chance to combine forces and pursue under-the-table prizes farther afield, a custom that continued between the wars. The village had a sign proclaiming it to be the 'Baseball Capital of Manitoba'.

Tournament play

Virden had a team in the Dennis County Baseball League in 1886 but locals saw the best baseball played by teams from the semi-professional exhibi-

> Gone are the open tournaments, the freelance pitchers, and the 'play for money'.
>
> Hal Duncan
> from *Baseball in Manitoba*

tion circuit of barnstorming women, blacks, odd combinations such as the House of David, or teams formed around disgraced members of the Chicago White Sox who had thrown the 1919 World Series.

Freestyle tournament play was the other significant venue of rural baseball. At a 1927 tournament Virden and Oak Lake were joined by teams from Saskatchewan and North Dakota as they competed for $1,000 in prize money. A year later 4000 Virden fans attended a local tournament featuring teams from Saskatchewan and Montana as well as two black teams, one of which, the Chicago Colored Giants, won the $600 first prize as well as a $200 side bet. By 1930 Virden were members of a Manitoba-Saskatchewan League but the real excitement was found in exhibition matches with teams like Gilkerson's Colored Giants. At a 1 July 1930 tournament in Souris, Virden beat the Brandon Greys in the semi-final and then thrashed Moosomin, Saskatchewan 21–1. Two homers were hit by Virden's Hap Felsch who had joined the team on 8 June. Felsch, the most notorious ex-White Sox member, had recently played, coached, and scouted in Saskatchewan. At Moosomin a week after the Souris tournament, Felsch's 14th inning home run gave Virden an 8–6 win in the semi-final. On 22 July at a Brandon tournament Felsch hit a triple and home run as Virden beat Gilkerson's Colored Giants 8–6. Felsch was gone by 1932 but Virden sponsored a post-season October visit by a team of major leaguers led by pitcher Robert 'Lefty' Grove who pitched two innings but refused to attend a post-game banquet.

Interesting characters abounded during the Depression as men unable to find regular employment forsook a more stable life and joined the ranks of the itinerant ballplayer. One such was Beano Melilli who pitched, played third base, short, first, and the outfield for the Brandon Greys over several seasons. His background was clouded. Different theories claimed he was a Hollywood actor, a contractor in Arizona, and a Mexican gold promoter. An accomplished musician, he was a popular drinking member of Brandon's sporting crowd.

In Carman, southwest of Winnipeg, tournament ball was the major focus of interest in the twenties

Swede Risberg, one of the eight Chicago White Sox accused of throwing the 1919 World Series, later played in western Canada

Hap Felsch, one of the eight Chicago White Sox accused of throwing the 1919 World Series, later played in western Canada. Felsch recommended a young Saskatchewan player, Archie Edwards, for a tryout with the Chicago Cubs in 1928. Homesickness got the better of Edwards and he returned home to pursue a teaching career. Of Felsch, Edwards later recalled, 'He'd do anything for a buck.'

and thirties. By 1949 the town was in the semi-professional Mandak League with teams from Manitoba and North Dakota. Among the recruits were four former negro league players whose league had collapsed following the integration of the majors, and a few locals like Almer McKerlie, known for his strong arm behind the plate.

Best symbolizing rural baseball was the Riverside Canucks team formed by a group of young farmers in 1946. Like most rural clubs they focused on tournament play. Spring seeding and fall harvesting weakened farm-based teams at either end of the season but in midsummer when there was time for community picnics and baseball, they were feared rivals. The Canucks won 20 of 23 tournaments in 1958. In 1961 however they joined the Manitoba Senior league as old-style tournament play declined along with other aspects of a once comfortable rural life.

Baseball in Brandon, Manitoba

Baseball in Brandon dates back to the 1870s where it survived till the end of the century on the largesse of local businessmen and eager athletes. The game's successful commercialization began with the building of a sports field in 1901 west of 18th Street and north of Victoria Avenue. It was later known as Kinsmen Park. The field was the setting for Brandon's first semi-professional team on which local talent was supplemented by outsiders. They joined the Manitoba Senior Baseball League in 1903. Other league teams included Portage La Prairie and Virden. The Brandon team played 16 league games along with a summer-long schedule of tournament play. The Brandon team lacked a nickname but possessed red-trimmed, blue uniforms with a large 'B' on the shirt. Ontarians continued to play a prominent role in provincial baseball. Brandon's squad included W. Babe (a former Ontario amateur star), Joe Cowan from Sarnia, George Dundas of Markdale, third baseman McAteer from Guelph, and Don Tiptoe of Woodstock. Two Americans and three Manitobans rounded out the line-up. Unfortunately Brandon's 7,000 residents did not support this commercial venture and the team disbanded before the end of the summer.

In 1908 a new Brandon team, along with Winnipeg, joined the six-team Northern League which was one of the first cross-border organizations in the west. American clubs included Fargo, Superior, Eau Claire and Duluth. Artie O'Dea managed league affairs out of Duluth and handled the finances of that city and Fargo. O'Dea's bookkeeping was more for his own benefit, and the league's collapse before season end left Brandon as champion.

The Brandon Angels surfaced in the all-Canadian Western Canada Baseball League in 1909. It remains the only Canadian league in organized baseball to have had teams from three Canadian provinces at the same time. Brandon's entry lasted through 1911. Perhaps their 1911 season highlight was the $400 draft price paid by the Chicago White Sox for Brandon's catcher.

Commerce had improved the quality of local play and increased spectator interest. The failure of three professional teams in Brandon within a decade suggested that the town was not yet ready for this level of baseball and so it experienced a temporary decline. The game rebounded in the twenties with the formation of the Brandon Greys, a squad made

up of equal parts local amateurs like Cliff Cory, Traves Fisher, Cliff Robinson, and Herb Stuart, and imported professionals from the United States, Mexico, and Cuba. This mixture of paid and unpaid players actually proved to be a viable long-term means of improving the quality of play. It maintained a financial stability that the previous all-professional teams lacked. The Greys played in tournaments all over the west and against barnstorming teams like the Kansas City Monarchs, the Toronto Oslers, and Grover Cleveland Alexander's Allstars.

After the Second World War the Greys joined the Mandak League along with Carman, the Buffalos and Elmwood Giants of Winnipeg, and Minot and Williston from North Dakota. The league included players like Gerry MacKay and Ian Lowe of the Brandon Greys, former black barnstormers, and even a one-time International League pitcher—Nuts Anderson from Rochester, New York. Hal Duncan recalls that a woman at a tournament at Indian Head (which attracted teams from California and the wonderfully named Carrot River Loggers from northeast Saskatchewan) in 1948 assumed that the Greys' name derived from the mixture of black and white players on the team.

In the early fifties minor league baseball was overextended and challenged by big league games on television. Though not a part of organized baseball, the Mandak League faced its own challenges. It considered dropping the two Winnipeg teams and forming a Class D circuit. Eventually the league disappeared and with it the Brandon Greys.

Manitoba's greatest baseball star: Russ Ford

In 1892 Russell Ford's family left Brandon, Manitoba. He had been born there nine years earlier, 25 April 1883, and so far as records indicate he had no further significant contact with either Brandon or Canada. It is a sign nevertheless of the dependent country's search for an identity that it finds it in native sons who gain glory in the country to which they are either dependent or within which they are but one of many regions.

The details of Russ Ford's early life are sketchy. His father Walter, who lived to 90, was an ace crick-

> Nothing can survive in Baseball which impairs the balance of the game.
>
> — Russell Ford in *The Sporting News*, April 1935

Russell Ford of Atlanta's Southern League team in 1906. It was here he discovered how to throw the 'emery ball'. Canadian Baseball Hall of Fame

eter and able to do handsprings on ice skates at the age of 40. His mother Ida was a second cousin to President Grover Cleveland. Russell's brother Gene was born in Milton, Nova Scotia on 16 April 1881. Sometime in the next two years the family joined the great migration to the Canadian west, likely travelling by way of the American Midwest through Chicago and St Paul, the CPR not as yet having connected eastern Canada and Manitoba. In Manitoba it was a period of transition from an earlier economy based on the fur trade and lumbering to the agricultural boom years from 1897 to 1910. Before the boom, however, the Fords had left for Minnesota.

Though baseball was played in Manitoba in the

1880s and the young Fords no doubt played some juvenile ball, Russ Ford later claimed,

> I learned most of the baseball fundamentals in grade school in Minneapolis, stuff like cutting throws from the outfield etc. My older brother Gene jumped from the sandlots to Indianapolis and also made the majors as a pitcher. After high school, I hoped to enter the University of Minnesota on an athletic scholarship, but was ruled ineligible for playing one game in the buckboard league at Enderline, North Dakota. I lost the game 2–1 and only learned after returning home that this game cost me my amateur status.

Manitobans considered Ford one of their own and perhaps that is only appropriate considering the itinerant nature of his life. Only in his last years did he truly settle down and that was in his wife's home town, Rockingham, North Carolina. Ford's first professional season was in Cedar Rapids of the Three-I League in 1905. Two years later he was 15–10 for the Atlanta Crackers who won the Southern League pennant. Here he developed the pitch for which he is remembered. The emery ball was an accident. On a rainy spring morning in Atlanta in 1908, a pitch sailed over the head of his battery mate Ed Sweeney and crashed into a concrete grating.

'Sorry Ed, that ball got away from me,' Ford yelled.

Ford's next pitch brought Sweeney to the mound.

'What did you do to the ball on that pitch, Russ?' Sweeney asked.

'Not a thing, Ed,' Ford replied. 'I just threw my regular fast one. Why are you asking?'

'You must have done something,' Sweeney countered. 'I never saw a ball act so queerly in my entire life. It seemed to jump sideways just as I was about to catch it. See if you can throw another one like it.'

Ford looked the ball over and noticed a small scuffed spot the size of a dime, where it had hit the concrete grating. He soon discovered that depending on how he held the ball it would break right or left, up or down.

'You can throw away your spitball,' Sweeney declared. 'This new pitch will win for you. But let's keep it quiet.'

Observers noted that the pitch broke even sharper than a curveball and because balls were seldom removed from play, once tampered with they had a long life. Ford's miracle pitch was seldom hit but a

Russell Ford with New York's American League team, 1910 (first row, far right). Man with hands on his shoulders is catcher Ed Sweeney, his emery ball co-conspirator. Canadian Baseball Hall of Fame

grounder became a defensive liability as simple tosses from shortstop to first often veered at the last second. He seldom used the pitch in 1908 but a year later, having been drafted by the New York Highlanders (later the Yankees) and sent to Jersey City, he roughened the ball's surface with a broken pop bottle.

'Things started to happen,' Ford recalled. 'The dizzy contortions that ball cut made me doubt my eyesight. I could get both the "hop" and the "sail" on the ball. The Jersey City players were missing my pitches anywhere from 12 to 18 inches. The success of the experiment gave me a big problem. This delivery could make me the greatest pitcher in the big leagues, but I had to keep it to myself.'

Ford knew that if his secret was revealed and others used the pitch the balance between offence and defence would be shattered and cries would go up to banish it. He discovered a clever ruse. Ford held the ball and his glove in such a way as to convince batters that he was throwing the then legal spitball.

His other problem was scuffing the ball. Obviously he couldn't take it back to the dugout. So he perfected a Rube Goldberg invention in which he strapped a piece of emery paper to his ring, cut a small hole in his glove and by tugging on a piece of

string was able to bring the emery into the hole from which he could rub the ball. His emery pitch went undetected because unlike most players who traditionally left their gloves on the field at the end of an inning, Ford took his wherever he went. Despite Jersey City's last place finish Ford used his new pitch to compile a 13–13 record, striking out 189 batters. Of course even a trick pitch can't be used all the time. Such pitches work because they start out looking like strikes but at the last minute fall out of the strike zone. If they were used too often, batters would simply stop swinging.

Called up by New York in 1910 he was an immediate sensation, winning 26 games followed by 22 in 1911. In his first start in 1910, he shut out Connie Mack's Philadelphia Athletics 1–0 and had nine strikeouts. In mid season he beat Cy Young who was pursuing his 500th win. Chief Albert Bender of the Athletics narrowly edged Ford out for the best pitching average. There was some controversy surrounding this result as Mack convinced New York manager Hal Chase to sit out Ford so the Athletics could rest Bender for the World Series. If Chase had allowed Ford to pitch, Mack would have had to start Bender to give him a shot at retaining his pitching average lead.

Ford claimed that the era's greatest star, Ty Cobb, once stole second and third base off him and then reversed his field going back to second and first. Flashing four fingers to the press box, Cobb is supposed to have yelled, 'Put that in your book.' Ford also recalled that Cobb once ignored an opportunity to spike the pitcher during a run down. 'Later I told him, "Ty, you could have hurt me and been excused for it." But he said, "Russ, I never hurt anybody who hasn't first tried to hurt me."' The greatest honour of Ford's career was selection to play in baseball's first unofficial all-star game in 1911 held in Cleveland to benefit the widow of the pitcher Addie Joss.

In 1913 the Highlanders trained in Bermuda. Returning to frigid New York Ford developed a sore arm. Nevertheless in mid season only an infield hit in the ninth inning broke up a no-hitter. In 1914 he was paid $24,000 to sign on for four years with Buffalo of the new Federal League. He won twenty games that season, but within two years was out of the major leagues playing in Denver of the Western League in 1916 and in Toledo in 1917. By now others had discovered his secret pitch, possibly from Ford's former backstop, Ed Sweeney. Not only pitchers but journeymen hitters and utility infielders all wanted a shot at the mound in hopes of reviving declining careers. In 1916 the emery ball was outlawed.

Ford's nomadic lifestyle continued after his playing days. He worked as a superintendent for the Submarine Boat Company in Newark during the last years of the First World War and coached college baseball in 1922 in Minneapolis. His family spent much of the twenties and thirties moving back and forth between New York City and Rockingham where Ford was alternately an assistant teller at the Bank of Pee Dee and a hotel clerk. He found work as a structural steel draftsman in New York in 1936 and the family lived there

Russ Ford had a sensational rookie year for the Yankees in 1910, with a 26–6 record and a 1.65 ERA. He was 21–11 the next year, but after that he began turning in losing records. By 1914 he was out of the league

until Ford's wife died in 1957. He returned to North Carolina and died in Rockingham in 1960. He was buried in the Episcopal Church by his longtime friend Reverend Howard Hartzell who called him a courtly gentleman with the manners of Lord Chesterfield.

In Manitoba he was recalled by writer Hal Duncan as that province's greatest baseball export, though how much he ever thought of his birthplace is not known.

CHAPTER SIX

British Columbia

Baseball and brass bands went together in the early stages of British Columbia's development. The band of the Royal Engineers performed in New Westminster from 1859 to 1863 as a militia, a fire brigade, a brass and a dance band. When the band was eventually reposted to Britain, several members remained behind. Early baseball is also found around the same time in New Westminster as American gold seekers descended on the Fraser River community. Bands also formed in Victoria in 1864, in Nanaimo in 1872 and Kamloops in 1886. Baseball play corresponds to these developments not only in Victoria in 1863, but in Nanaimo by the 1880s, and in Kamloops in 1885 during construction of the Canadian Pacific Railway. The dates are not coincidental. They demonstrate that settlers sought the amenities of civic life almost from the moment of arrival and in so doing brought with them the attitudes and pleasures of past lives and places.

Just as new technologies and forms of government take years to develop in their home base but are adopted seemingly overnight in new territories so it is with new recreations and cultural ideas. The evolution of the sport which in the east happened over many years had a different history in the west where time was compressed. The baseball tradition arrived in the early 1860s in an at least partially developed form and was subject to new influences on an annual basis. Within a decade international matches were being played.

A San Francisco beginning

British Columbia's baseball story has its beginning in San Francisco with the arrival on 10 August 1849 of Alexander Cartwright, the generally acknowledged father of modern baseball and a member of baseball's pioneering club, the New York Knickerbockers. Even though he had spent five months travelling from New York to California his stay in the bay area would be only five days. During that time he met his brother Alfred and another former Knickerbocker, Frank Turk. We can speculate that baseball news from the east was part of their conversation and perhaps they played a game or two. Alexander then left for Hawaii and though he had plans to go on from there, his aversion to sea travel was such that he never again left the island. There seems little doubt however that his brother Alfred and friend Frank played some role in popularizing the game on the coast. Throughout the 1850s the game was popular in California, long before it caught on in points between the American east and the Pacific coast. A formal club, the Eagles, was founded in 1859 and the sport flourished as an increasing number of migrating easterners settled in San Francisco. By 1867 the *Pacific Base Ball Guide* speculated that over a hundred clubs existed throughout the state from Santa Clara to San Jose.

Up the coast from San Francisco, a small outpost of the Governor and Company of Adventurers of England Trading into Hudson's Bay, known to us as Victoria, had perhaps 300 citizens. One of these was an Englishman, Alfred Waddington, who arrived in the spring of 1858 to open a branch of a San Francisco wholesale grocery firm. By the end of the next year he had been followed by 20,000 others, many of whom paid $15 for the steamer trip from San Francisco to

Victoria. In the fall of 1858 gold had been discovered on the Fraser River. As in California a decade before and as was to occur in the Yukon at the turn of the century, there was a mad rush to the site.

Victoria was the jumping-off point to the Fraser and of the 450 who first arrived in April 1859 aboard the American steamer *Commodore*, 60 were British subjects, an equal number were white Americans, 35 were blacks fleeing discrimination in California, and the rest were formerly from Germany, France and Italy. To the British governor Douglas it signalled the arrival of a significant 'anti-British element'.

Most arrivals were temporary residents but at its peak Victoria's population swelled to 7,000. Its social life was transformed and the new colony of British Columbia was established before the end of the year. This British outpost with its attendant customs lived in uneasy tension with a polyphony of competing cultures of which the American emerged as dominant.

Bat and ball games in Victoria

Bat and ball games provided a forum of conflicting tastes. Victoria's first cricket club had been formed in September 1858. Thomas Harris, later Victoria's first mayor, was club president and games were played at Beacon Hill and Colwood. According to historian Geoff LaCasse the first record of a baseball game

appears in March 1863. It was played on the cricket pitches of Beacon Hill in the middle of the horse-racing track. It seems certain that baseball was introduced to Victoria by ballplaying gold seekers from California, some of whom may even have been former members of the first California clubs influenced by the Cartwrights and Frank Turk. As is the case in most of North America unrecorded informal games would have preceded actual written accounts by several years. This conclusion is at least partially confirmed by mention of a baseball game in an advertisement for New Westminster's 24 May 1862 Queen's Birthday celebration. New Westminster was closer to the actual gold rush location. Once there, prospectors would have had more time and opportunity to play the game thus explaining its formal mention at an earlier date than in Victoria. Games were played primarily in March/April and September/October when claims were not worked and the weather was good.

Victoria's first team, the Olympics, is mentioned in 1866 alongside the City Base Ball Club. The Olympics eventually became the Amity Base Ball Club, the most important early baseball team in the province and one which according to LaCasse spanned the period from 'old guard to new, amateur to professional, open fields to enclosed parks, middle class businessmen to professional players'. The final piece of this early baseball puzzle was provided in

A baseball game in Victoria, British Columbia. British Columbia Provincial Archives, 39777

1872 with the visit to Victoria of San Francisco's first team, the Eagles.

On the west coast through the 1870s American influence and the popularity of baseball grew. Petitions were occasionally circulated promoting American annexation of the colony but in July 1871, with the promise of a railway to the west coast, British Columbia became Canada's sixth province. For the next 15 years, however, it would remain isolated from the rest of the country, allowing baseball's strong north-south orientation to become even more firmly rooted. Baseball's only significant sporting competitor was cricket which never really attracted interest beyond its core of hardy anglophiles.

Baseball follows the railroad

Baseball growth followed the advance of the railroad. Although the CPR created a strong rival in lacrosse, enabling British Columbia teams to compete in national championships, it also brought a new legion of baseball players and supporters such as the Vancouver entrepreneur W.D. Haywood who had worked on the Colorado railroads in the 1870s.

Vancouver had been an insignificant port city of 1,000 in 1886 but its selection as western terminus of the CPR spurred its growth to a population of 13,000 within five years. One of the first signs of Vancouver's new prosperity was the importing of players by rail to play for the Vancouver Terminals baseball club.

The railroad also influenced the growth of baseball throughout the British Columbia interior. The community of Donald near the province's eastern boundary sprang up overnight, at first as a construction site and later as a station. Railworkers played the game as early as 1883 but Donald baseball's great year was 1888 by which time the line had reached the coast. In that year Donald entered a Dominion Day tournament organized by the St Andrews Caledonian Society in Kamloops. Play was haphazard and primitive. Donald destroyed Nicola 32–0 and Kamloops 40–0. Its gambling supporters were ecstatic. Adult play in these settings resembled children's sandlot games of today in which gloves might be missing and only one team has a few good players.

The railway not only brought better players to the region but it also allowed regionally based tour-

Baseball playing at Kaslo in the early stages of British Columbia's development. British Columbia Provincial Archives, 1752

naments which countered the north-south orientation of most play. Donald went to a Calgary tournament and were narrowly defeated in the finals by
Medicine Hat 21–20. A second Kamloops tournament was scheduled for September and was promoted as the provincial championship. Donald brought
in several Minneapolis professionals including a
pitcher—March, who was paid $400, a catcher—
Gallagher, and a shortstop—Rathburne from Manitoba. This is an indication of the quality of play in
such games for those three positions are usually taken
by the best players because they receive most of the
defensive action. Donald lost to the eventual champions, the Amity club from Victoria, 10–6 and finished
behind the runners-up from New Westminster.
Donald ended their season by defeating their old
nemesis Medicine Hat but the club did not reappear
a year later. Employment on the rail line declined, the
town's only newspaper relocated to New Westminster,
and its ballplayers were scooped up by Kamloops.
Soon the community itself would virtually disappear.

Baseball in Kamloops

Baseball—and rail construction—had reached
Kamloops, 150 kilometres west of Donald, in 1885.
Before the CPR arrived, Kamloops had functioned as
a trading centre in the British Columbia interior. The
railway brought a new vitality and its workers largely made up the rosters of Kamloops' two teams—the
Blues, and the Knickerbockers, whose executive
included a future mayor and a fire marshal. These
workers followed construction to the coast and both
teams dissolved in 1886. The rail line eventually
brought an influx of permanent workers, many of
them eastern Canadians who had either played or
watched the game and in 1888 a Kamloops club and
a CPR team formed. They shared practice space on the
grounds behind the Robson and Lee store and
helped sponsor the 1 July tournament, won by
Donald, and September's provincial gathering with
its $300 prize, won by Amity. In his short story 'The
Elks Come to Town' (published in a 1987 issue of
Descant magazine) poet George Bowering incorporated a New Year's Day 1889 game between these two
Kamloops teams which was delayed by an eclipse of
the sun.

The following year the Kamloops teams united
and brought in seven members of the semi-professional Minneapolis Lyndales including Donald's for-

mer pitcher March, third baseman John Hirchner
and shortstop Joe McCrum. The team had only two
local players. The team rode the CPR to the coast and
easily won two exhibition games from Amity. The
Kamloops players were severely criticized by
Victoria's newspapers for their poor sportsmanship
which included arguing with umpires, taunting their
rivals, questionable on-field tactics, and casual arro-
gance in placing bets on themselves to win. In a
game against Port Townsend they not only bet
$1,000 on victory but guaranteed a shut-out.
'Baseball as played by Kamloops,' historian LaCasse
notes, 'more closely resembled the professional
leagues back east then the gentlemanly game in
British Columbia.'

At the 1889 Kamloops tournament, the professional home team narrowly beat Amity 5–4 in extra
innings in the open competition. The two teams met
later in the same tournament in what was supposedly a more local championship. In this game Amity
closed to within 6–5 of Kamloops following a wild
throw to first base with the bases loaded. After the
Kamloops players harassed the umpire, he reversed
the original call and called the Amity players out.
The score reverted to 6–2. Amity immediately left
the field in protest.

The provincial supremacy of a team from British
Columbia's interior was temporary. Coast teams
refused to play Kamloops and that team abruptly disbanded, its unpaid players threatening legal action.
Most of them soon departed but McCrum and
Hirchner stayed and found jobs with the CPR. In
1946 the local Inland Sentinel newspaper would eulo-
gize McCrum, who died at the age of 82, as one of
the pioneers of the area.

Vancouver faces the Orient

British Columbia's best baseball would be found
henceforth in Victoria and Vancouver, though the
smaller centres have been home to a majority of the
province's major league ballplayers. As the gateway to
the Orient, the west coast of Canada and particularly
Vancouver is caught in the social contradictions of
that location. Captain Cook brought British influence
to the region in 1778, four years after Spanish ships
had patrolled the coast. In 1792 George Vancouver
mapped the coast from present-day Oregon to Alaska.
The native population, whose occupation dates back
6,000 to 8,000 years or just after the last ice age, were

The 1903 Amity Club of Victoria. Hal Chase (far right, top row) was to be implicated in several gambling scandals during his future major league career. British Columbia Sports Hall of Fame and Museum

Ballplayers (Dominic Nicholas second from left) and two Kootenay native women at Lake Windermere, British Columbia, 1922. Glenbow Archives, NA 1135-35

considered only in so far as they contributed to the fur trade economy of the British.

The Gold Rush, of course, forever ended the region's relative isolation from the larger world. Among the earliest arrivals with the California gold seekers were Chinese workers. Their numbers multiplied with the demand for labourers on the construction of the Canadian Pacific Railway in the late 1880s. In whatever occupation they chose the Chinese were resented for their willingness to work long hours at low wages. Most of them were single men whose only desire was to bring over their families from China. Within the white community one of the earliest organizations of working people, the Knights of Labour, was formed to combat 'renewal or increase of Mongolian competition to citizen labour'. The anti-Chinese riots of 1877 were the first overt expression of white antipathy to Chinese labour practices. The first Japanese immigrant to the region was Manzo Nagano who settled in Victoria in 1877. Japanese arrivals found their niche in farming, the pulp and paper industry and as fishing boat operators, an occupation many had pursued in Japan.

As Vancouver grew after 1886 it quickly became the Canadian focal point for contact between the rest of the country and the nations of the Pacific. In 1891 Sir William Van Horne who had managed the CPR construction launched the Empress line of Pacific steamships, which linked Vancouver and Hong Kong. In a few years these Pacific linkages would facilitate the first trip of a Canadian baseball team to Japan.

Vancouver baseball

Baseball in Vancouver mirrors the city's period of growth. In 1889 the *News Advertiser* claimed that baseball had first been played in Vancouver two years earlier. Given the popularity of the game in other parts of the province, particularly Victoria, it's likely the game was played before 1887. It would nevertheless have been an ill-formed sport, and one not played very well. Vancouver's team was easily defeated at the 1888 Kamloops tournament. To improve the local calibre of baseball the Vancouver Terminals began in 1889 to import eastern talent by CPR.

In the 1890s the game struggled for recognition and suffered economic problems in the decade's middle years. One of the city's best pitchers was forced to leave Vancouver in 1895 for a job in the Orient. Victoria's 30-year civic headstart on Vancouver

worked to the advantage of its ballplayers and teams, and the opening of a rail connection between Vancouver and Seattle in 1892 allowed Victoria in 1896 to join a league with teams in Seattle, Portland, and Tacoma.

Vancouver's baseball evolution was encouraged by the leadership of several baseball-loving Americans. In 1889 the American consul Bolton was named honorary president of the Terminals club, and in 1900 another American consul, Dudley, allowed his office to be used for meetings of the Vancouver baseball club. He was appointed their honorary vice president. By the turn of the century Vancouver's teams were asserting themselves within British Columbia thanks to local leadership, improved facilities, and the city's growing population. In 1904 Vancouver had 40,000 citizens, a total that doubled within five years. Baseball leadership was provided by, among others, William 'Sweet Billy' Holmes, a longtime Vancouverite, who played junior baseball in 1890 and by 1903 was club secretary for Vancouver's best team.

Vancouver's first baseball ground, on Cambie Street in 1887, was on CPR land leased to the city. Here, says LaCasse, Al Larwill, an early lacrosse and baseball player (born in 1849) lived in a shack by himself and 'taught successive generations of schoolboys the rudiments of the game and morals of the world'. The grounds were not maintained, however, and eventually baseball relocated to the somewhat inaccessible Brockton Point Grounds, a factor in the game's commercial decline.

Baseball next moved in 1900 to the Powell Street Grounds in an area bound by Cordova, Dunlevy, Jackson, and Powell Streets. At the enclosed Powell Street Grounds spectators paid a 25-cent admission. The site was developed by a consortium which included local businessman A.E. Tulk, and its commercial possibilities allowed Vancouver to have its own team of semi-pros within two years. Vancouver was visited by a succession of teams from the American northwest and California, as well as the Boston Bloomer Girls. Baseball in the city had clearly entered a period when commercialism would guide the sport's future.

In 1901 organizers from Vancouver, Victoria, New Westminster, and Nanaimo formed the British Columbia Amateur League. Vancouver's commercial success at the box office allowed it to hire better players, often by the ruse of finding them 'jobs' with local businesses. The team compiled a 5–1 record in 1901 and dominated the league. Smaller teams and local newspapers complained that Vancouver's hiring practices displaced local talent and upset the balance of power, and the league folded after the 1903 season. In 1905 Vancouver and Victoria were founding members of the professional Northwestern League. The Powell Street Grounds had been taken over by the city so Vancouver's top pro team now played in the new Recreation Park grounds. With the exception of the period between 1923 to 1936, and in the 1970s, Vancouver has had a succession of minor league teams made up of imported professionals.

Amateur ball in Vancouver and the Japanese community

Below this level Vancouver has had a succession of locally sponsored teams and leagues. In name most were amateur, but under-the-table payments and work placements which characterize all forms of commercial sport, especially at the top level of senior baseball, were common. Between the wars over 50 senior teams in British Columbia played in nine leagues of which three were on Vancouver Island and six on the mainland including Vancouver's Terminal League. The champions of each league participated at season end in a series of playoffs to determine a provincial champion. These leagues were a reasonable goal for those playing in junior and juvenile competition which had developed by this time.

One such Terminal League team was the Asahis, representing Vancouver's Japanese Canadian community. Japanese Canadian children meanwhile played for the Young Clovers, the Beavers and the Athletics. Around the turn of the century thousands of Japanese, some stranded in Hawaii because of a sudden cutoff by United States Immigration, had come to British Columbia. Many settled in an area around Powell Street which had become known as Little Tokyo. By 1907 the 500-member Asiatic Exclusion League passed a resolution against the admission of Japanese arguing that they kept salaries artificially low because of their willingness to accept 'coolie wages'. The motion was endorsed by all political parties.

On 8 September that year a protest march turned into a riot. In Chinatown the rioters met little resistance but in Little Tokyo residents had armed themselves and drove out the invaders. Prime Minister Wilfrid Laurier dispatched a future Canadian leader William Lyon Mackenzie King to investigate the situation. Payments were made to many affected Oriental immigrants though King rejected the claim of two Chinese opium makers. The Japanese government responded to Canadian concerns by setting an emigration limit.

Chinese Canadians reacted to these indignities by retreating further into their own lifestyle. Japanese Canadians showed a far greater interest in the customs of their new homeland. In the first decade of the century they had easy access to the Powell Street Grounds. The Nippon Baseball Club was formed around 1908. It was made up of young Nisei who had played the game informally in the shadows of the ballpark and others like Kunitaro Noguchi who had played baseball in Japan.

Baseball had been introduced into that country in the 1870s but remained a novelty until Hiroshi Hiraoka, who had studied in the United States and returned to become a supervisor for the Japanese Ministry of Engineering, organized railway employees into Japan's first baseball team. It was a moment of supreme national celebration when in 1896 Japanese students at Ichiko, one of the five élite prep schools for the national university, defeated a team of Americans from the Yokohama Athletic Club. Still, baseball in Japan was largely a scholar's game and the majority of Japanese working people who emigrated to North America had little awareness of it. Their exposure to baseball and that of their children was in America.

When the Nippons of Vancouver played a Victoria Nippon nine at the Powell Street Grounds it marked the first time that the white population of Vancouver took a positive interest in their Japanese neighbours. For many young Japanese it overcame the loneliness of life in Canada. While there were about 10,000 Japanese in Canada by 1914 they had until recently been mostly young men. Ballplayers along with teachers at the Japanese Language School and newspaper workers were among the first members of the Danwakai club, a discussion group for the exchange of opinion. Other ballplayers gathered at a Japanese tavern to drink and talk about the day's game. Throughout the province Japanese Canadians played baseball in Ocean Falls, the Okanagan, and in the logging camps in the interior of the Fraser Valley. In these places as in Vancouver baseball created civic pride in the Japanese community and a bond with the larger community.

The rise of the Asahis

The Asahi team first took shape in 1914 under the guidance of Matsujiro Miyasaki, owner of a clothing and food store at 200 Powell Street. His players included four youths with connections to the Japanese village of Kaide-ema, Hosaka. These were the three Kitigawa brothers, Hatsu (Mickey), Yo Horii, and Canadian-born Eddie, as well as Yosomatsu Nishizaki. Miyasaki's star player, Canadian-born Tom Matoba, was an exception to the Asahi strategy based on singles, stolen bases and the hit and run. He favoured the long ball. The young Asahis took on everyone including longshoremen, firefighters, and teams from up the shore or in the interior. No other Japanese Canadian team could match them, and they became even stronger in 1918 when the older Nippon team folded and their best players joined the Asahis, who were now the community's leading senior team.

By 1919 the Asahis had won a local title. They were so good that the team split in two for the 1921 season. One toured Japan with Dr Saitaro Nomura and the other played locally under the Tigers banner. In a successful attempt to cement bonds with those outside their community, the Asahi management invited four non-Japanese Senior League players to join 12 Asahis on the Japanese tour. The barnstorming team boarded the *Kashima Maru* out of Vancouver in late August for the 14-day boat ride. Their possessions included baseball equipment and a film on Canadian industrial development provided by the British Columbia government. On arriving they discovered that a Japanese American team from Seattle was also on tour. Both teams played against a series of Japanese university teams and funds were used for the establishment of a Japanese baseball college founded by an American naval officer.

In the early 1920s, Harry Miyasaki dreamed of creating a championship calibre team to win the

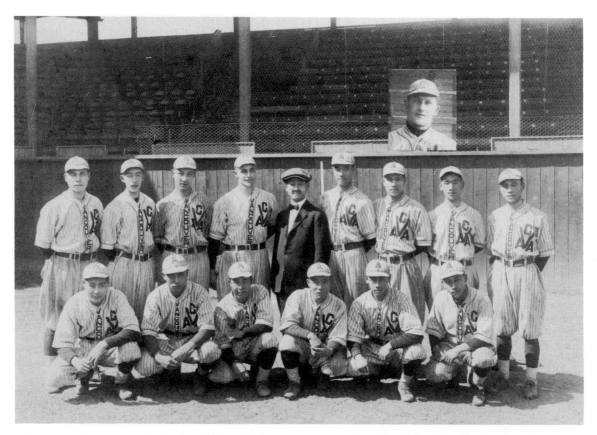

A Vancouver all-star team, of Asahi team members and selected non-Japanese local ballplayers, who toured Japan in 1921. 'It took us 14 days to get there by boat and I was seasick the last 11 days,' said Ed Kitagawa. Back (L-R) Joe Nimi, Ted Furumoto, Harry Yoshioka, Ernie Papke, Asahi president Dr Nomura, Yuji Uchiyama, unknown, Mickey Kitagawa, Tom Miyata; front (L-R) Yo Nishizaki, Yoshi Oka, unknown, unknown, Yo Horii and Ed Kitagawa; insert: Joe Brown

Terminal League title against the bigger, harder hitting Occidental teams. With players like Junji Ito, Roy Yamamura and Tom Matoba, Miyasaki built a team based on speed and sparkling defence. It was not unusual for the Asahis to score runners from second and third on a squeeze play, and their mastery of the cutoff prevented other teams from copying their patented double steal. In 1926 this team realized their dream by winning the Terminal title.

Junji Ito, who had been educated at an agricultural college in Japan, emigrated to Canada and first played in the Vancouver City Baseball League in 1920. He earned the title 'King of Bunting'. Even though the opposition knew what he would do at bat, his average climbed to over .400. He taught an entire generation of Japanese Canadians the skills of his craft. Roy Yamamura, barely five feet in height, was dubbed 'the dancing shortstop' for his spectacular fielding. And on the basepaths, his lightning quick instincts electrified spectators. In 1925 he stole over 50 bases. He is perhaps best known as the only Japanese Canadian to play for the Arrows team in Vancouver's top division for two years.

Within the Japanese Canadian community the team was a powerful symbol of their participation in the larger society. Many years later 94-year-old Iwaichi Kawashiri recalled:

In 1928, I lived on Powell Street, running a rooming house. My greatest pleasure in those days was to grab 10 cents and head for the ballpark to see the Asahis play. That was the beginning of my love for baseball. Powell Grounds was their stamping ground. . . . During those years the Japanese people were discriminated against at every turn. To have a Japanese language school to teach our children the Japanese language [and] culture was highly suspect. We were constantly criticized for congregating in one area or working too hard to eke out a living. . . But the barriers came down whenever the Asahis played baseball. Naturally, there were Japanese fans. But it was the applause from the Occidental fans which made us so proud. There was one coach on one of the opposing teams named Don Stewart. When his team was at bat, some of the Occidental fans would encourage the Asahis to execute a double play. This made Don Stewart so angry, he became livid and grabbed the wire fence and yelled at the fans to shut up. 'You root for Japs,' Stewart would holler. I felt good.'

A grudging respect emerged out of these contacts between Japanese Canadians and the largely Occidental population of Vancouver. In winter Yo Horii even formed an all Japanese hockey team. In

The 1926 Asahis, champions of Vancouver's Terminal League: back (L-R) Ed Kitagawa, Sally Nakamura, Reggie Yasui, Junji Ito, Tom Miyata, Mickey Sato, Tom Matoba, Charlie Tanaka, manager Harry Miyasaki (the John McGraw of Japanese Canadian baseball); front George Kato, Ty Suga, Roy Yamamura, Roy Nishidera, Frank Nakamura, Yo Horii Kitigawa

the background, however, disturbing voices were to be heard. In 1925 the British Columbia provincial legislature debated a ban on the newly arrived Ku Klux Klan. One member said that if the Klan was against Catholics, Jews, and Negroes there was no place for it in a British community, but 'if it freed the province of orientals' he was for it. Nevertheless in the 1930s when public opinion sided with the Chinese in their war against the invading Japanese, few in Vancouver had any bitterness towards Japanese Canadians because that community had more successfully integrated into the west coast lifestyle than had local Chinese.

For their part Japanese Canadian ballplayers were eager followers of baseball matters in North America. Kenichi Suga, a southpaw pitcher, was nicknamed Ty in honour of the legendary Cobb. And there was great excitement in the community when Babe Ruth played an exhibition game in Vancouver in the 1930s while on his way to Japan. In 1935 Japan's greatest team, the Tokyo Giants, toured North America, playing the Asahis in Vancouver on May 24 at Con Jones Park and travelling as far east as Saskatoon. Here on 19 June the Tokyo all-stars led by two pitching legends of Japanese baseball, Eiji Sawamura and the Russian speedballer Victor Starffin, beat the Saskatoon All-Stars 14–0.

As the Second World War approached the good relations created by Canadian based teams like the Asahis or touring teams like the Tokyo Giants were lost in the din of world affairs. The Asahis were swept up by these forces. Canadian-born Ken Nakanishi pitched for the 1933 Terminal League championship team. On travelling to Japan shortly thereafter he was conscripted and not heard from again until 1938 when it was learned that he was being treated at an army hospital in Tientsin. He was expected to return to the front when his leg healed. Friends in Vancouver sent him a leather coat with messages and signatures attached to the lining. Frank Ejima, another Canadian-born Asahi, went to Japan in 1934 in the depths of the Depression looking for work which he eventually found in Shanghai, as an interpreter for the Japanese army. Abie Korenga, a University of British Columbia graduate and member of a 1938 championship team, was later reported missing in the Malay jungle while working as a Japanese army interpreter. He too had been conscripted while visiting Japan. Ken Noda, a Japanese national and star pitcher for the Asahis in 1934, later

played ball in Seattle before returning to Japan. He was killed in action in China.

The Asahis played through the early stages of World War Two but the Japanese government's attack on Pearl Harbor on 7 December 1941 shattered the fragile bond between Japanese Canadians and the larger community. Possessions like fishing boats, businesses, and homes were confiscated. In Vancouver 8,600 Japanese Canadians were expelled to inland internment camps. George Yoshinaka pitched for the Asahis in 1940. During the war he was interned in New Denver, British Columbia where he played baseball against teams from Slocan, Lemon Creek, and Kaslo. Ty Suga coached the Lemon Creek All-Stars to the Slocan Valley championship in 1942 and 1943 in a league which included Japanese Canadians and Occidentals. After the war it was proposed that interned Japanese Canadians be deported to Japan regardless of their birthplace. In response many chose to leave their tight-knit little community and settle in other parts of Canada. Never again could they be so easily picked out for persecution. Yoshinaka moved to Alberta where he coached some championship amateur teams in the 1960s. In 1975 he was part of a group of ten investors who obtained a class "A" Pioneer League franchise, the Lethbridge Expos, featuring a future star—Andre Dawson.

Professional baseball in Vancouver

Within the larger community the best baseball in Vancouver in the 20th century was played either in the many senior leagues or the professional teams which represented the city in a variety of minor leagues. The two most significant promoters, Bob Brown and Nat Bailey, were both American-born. 'Ruby Robert' Brown from South Dakota, a former Notre Dame football player, had purchased the professional Beavers for $8000 in 1910 after the team had struggled financially for several years since its establishment in 1905 by local entrepreneurs A.E. Tulk and A.J. Mayo. Brown's teams won titles in 1911, 1913, and 1914 but the First World War ultimately helped kill the Beavers' league, the Northwestern.

Around this time Brown built Athletic Park, and often used his own pension funds from his days as a Roughrider in the Spanish American War (though he never actually saw service having been stricken with dysentery while in camp in the American south) to sponsor senior baseball. He imported American col-

Bob Brown, often dubbed the father of British Columbia baseball. British Columbia Sports Hall of Fame and Museum

lege players from Washington and Oregon to improve the quality of play. One of those hustling concessions at Athletic Park in 1918 was Nat Bailey who had been born in St Paul, Minnesota and came to Vancouver by way of Seattle. He said Brown would very nearly cry when soccer players tore up his field with their cleats. Bailey recalled the best year ever at the Park was 1933. 'Admission was 25 cents and women were admitted for free. We were bringing in 3,500 to 5,000 every night of the week. Selling peanuts for five cents a bag we used to go through four or five thousand pounds.'

The Capilanos brought organized baseball back to Vancouver in 1937 and except for a brief respite during the Second World War the Western International League team lasted through 1954. Significantly Capilano Stadium opened in 1951 a monument of sorts to Brown who became president of the League in 1953. Bailey was now the prominent organizer on the Vancouver scene. He led the Vancouver Mounties into the Pacific Coast League in 1956 but the team was dogged by financial woes partially because Vancouver's Sunday 6 p.m. curfew prevented them from playing doubleheaders which always drew large crowds. The team ignored the law and were charged. President Bailey told the court that 'it would be impossible to operate without the big turnouts for Sunday games.' Nevertheless the Mounties were fined $50 on each of three counts. The team survived until 1969, a period in which baseball was in decline throughout all of North America.

Vancouver returned to the Pacific Coast League in 1978 and Capilano Stadium was renamed in hon-

Opening day for the Vancouver Beavers, April 1915. Pennants commemorate 1913 and 1914 Northwestern League pennants. British Columbia Sports Hall of Fame and Museum

our of the recently deceased Nat Bailey. In its first several years the team was owned by a succession of mercurial west coast sports entrepreneurs: Harry Ornest, Nelson Skalbania, Jimmy Pattison. Molson Breweries and Japanese interests have been recent owners. But the club's most distinctive feature is its ballpark which writer Tom Hawthorn described as 'snuggled in the lee of Little Mountain in the heart of Vancouver. A tree-covered slope rises just beyond the outfield fence. . . . The brush and towering firs of Little Mountain remain untouched, an oasis of nature surrounded on all sides by the city.' The site is so wonderful that the team opted to stay here even after Vancouver built Canada's first domed stadium.

Besides its name Nat Bailey Park retained at least one other powerful connection to its city's baseball heritage. Foul ball netting stretches from first to third base and upwards from the first row of seats to the Park's rooftop. It not only protects fans from flyballs but retrieves them for the team's later use in batting practice. Unravelling the net in spring and rolling it up again at season end is a more complicated task than it might appear. In 1992 four Japanese Canadian fishermen began their 14th season overseeing the job. One of them, Terry Nakatsu, had played for the Steveston Fujis in a community by the Fraser River and was invited to play for the Asahis. Though interned during World War Two he played baseball for a sawmill team against teams from Kelowna and a local Doukhobor community. While many others left British Columbia after the war, Nakatsu remained to coach an industrial league team at the Powell Street Grounds and prepare for his eventual call to Nat Bailey Park.

Postscript:

Kingston man popularizes the game in the Yukon

Robert J. Eilbeck moved to Kingston in 1871 to become an agent of the Dominion Telegraph Company on Ontario Street. He helped organize a local team, the St Lawrence and played first base and batted fourth in their game against Cape Vincent, New York. For much of the decade the Kingston St Lawrence were a leading team in Ontario, occasionally recruiting American professionals like Rafferty and Dygert. In 1875 Kingston beat the reigning Canadian

champion from Guelph and in 1876 the Eilbeck-led team that included five American pros joined the first ever Canadian baseball league. Eilbeck's sporting interests broadened in the 1880s. He played cricket and captained the rowing club as a member of which he watched the great Hanlan-Ross race at Ogdensburg, New York.

In 1884 'the old man' helped revive the local baseball team and in 1888 he managed Kingston's entry in the Eastern International League. His wife had died in 1881 and by the 1890s his travels took him farther afield. In 1892 it was reported that he had arrived home from Helena, Montana to spend Christmas with his family. He was now working as a mining engineer. So it wasn't surprising that he joined the Gold Rush trek to the Yukon in the late 1890s.

Eilbeck was prominent in local sports and arranged railway tickets for the Dawson City hockey club that challenged for the Stanley Cup in 1904. Baseball had arrived in the Yukon at the time of the Gold Rush and Eilbeck was a key organizer. His chief protagonist in Dawson was Charlie Lamb, a well known sportsman and miner who was forever 'throwing down the gauntlet' to Dawson's constabulary. As befitted an unstable community baseball matches were often rowdy affairs. A 1903 match between the

The champion 'Arctic Brotherhood' baseball team in Dawson, 1900. Vancouver Public Library Collection

Game in the Yukon. Whitehorse team on field and North-West Mounted Police buildings in background. MacBride Museum Collection, Yukon Archives

Idyle Hour and Civil Service teams had to be broken up by the police after a fight erupted between one Captain Bennett and an Eldorado Smith.

In the summer of 1904 gamblers sponsored an international series between a team from Skagway, Alaska and the Yukon town of Whitehorse. Led by the legendary slugger Jack Keating, known from Seattle to Dawson, Whitehorse surprised the favoured Alaskans 10–9 on the Queen's Birthday. At the return match in

Skagway the Yukoners completed the sweep with an 8–5 victory on the 4 July holiday. Whitehorse fans mocked Skagway as being too windy, and Juneau and Douglas as too wet, to offer serious competition. When Eilbeck returned to Kingston in 1905 he was able to tell everyone that Yukoners had gone wild over baseball. He was anxious to get back home, he said, as the baseball season would soon be over and his son J.M. Eilbeck was coaching the Yukon team in his absence.

CHAPTER SEVEN

Québec

Québécois and early baseball

Georges Maranda had what is referred to as a mere cup of coffee in the major leagues with San Francisco and Minnesota in the early 1960s. He was described as the French Canadian with the Charles Boyer voice. His family had lived in Quebec since 1681 when Jean and Jeanne Maranda and a family which eventually totalled 25 children arrived in the French colony.

Georges Maranda from Lévis, Quebec. Canadian Baseball Hall of Fame and Museum

Family members had played baseball for the Club Lévis since 1934 and Maranda was signed by a Quebec-born former major leaguer, Roland Gladu, to a Boston Braves contract in 1951, after pitching a no-hitter for Lévis in a Quebec Senior League game.

His baseball career took him from Lévis, Quebec, to Panama, Mexico, the Dominican Republic, Hawaii, and Japan. Teammates with Louisville in 1959 included fellow Quebeckers Ron Piche and Claude Raymond. Among his varied post-career jobs was that in public relations for the Disco Bar le Vieux Chêne. The Lévis ballpark that opened in 1973 was named Stade de baseball Georges Maranda.

The Maranda family's deep heritage in the province and Georges' baseball career, like that of others ranging from Vadeboncoeur to Boucher, demonstrates the significance of baseball within the life of the French Canadian community of Quebec. The history of Quebec baseball corresponds to larger forces at work within the French-speaking community.

The Bouchers

This story begins with the rather unusual connection between real life and fiction in the story of the two Denis Bouchers. The first Boucher is a gifted Québécois pitcher signed from under the noses of the Montreal Expos by their Canadian rivals, the Blue Jays. Boucher rises quickly in the Toronto farm system as French Canadian baseball fans seethe at the Expos' failure to develop a local baseball hero. He starts the 1991 season with the Jays but despite press reports comparing his style favourably with that of

Jimmy Key he has only limited success and his name goes almost unnoticed in a mid season trade with Cleveland. Flash ahead two years and several other playing stops, and he is starting for the Ottawa Lynx of the International League, one step away from promotion to the big leagues. His eventual arrival in Ottawa's major league affiliate is greeted with grand enthusiasm and huge crowds. Denis Boucher has finally made it to Olympic Stadium in the uniform of the Montreal Expos. Before the season's end he has won three games for them, including one game in which he teams with catcher Joe Siddall from Windsor, Ontario, only the fifth all-Canadian battery in major league history.

The second Denis Boucher is a budding young reporter in Roger Lemelin's *The Plouffe Family* (1948), a novel that will have an enormous following when converted into a Quebec television series in the fifties. Boucher befriends an American Protestant minister, Tom Brown, who in turn has shown interest in the baseball future of Guillaume Plouffe. The local Catholic priest Folbeche is concerned not only about Boucher's fraternizing with Brown but about the game of baseball which the priest views as one more attempt by outsiders to weaken the fabric of his French-speaking community. Boucher tries to sway the priest by warning him that while Moses was conversing with Jehovah his people were building a golden calf. 'Today your parishioners' golden calf is baseball. Don't leave them to their own devices,' Boucher warns.

Boucher's plan requires Tom Brown to swing helplessly at three ceremonial pitches from the priest prior to an important exhibition game. The big day arrives and the priest who has never played before tosses two clumsy pitches at the American. Brown deliberately misses but, his pride suddenly piqued, he resolves to hit the third pitch. The priest's quick throw is over the plate for strike three before he can react. In the translated version of Lemelin's story, a band of men hoist the priest on their shoulders chanting, 'He has won his epaulettes . . .'—the same action and words a real-life crowd had chanted in 1946 after Jackie Robinson had led the Montreal Royals to the minor league baseball title. Lemelin's portrait of working-class life in Quebec acknowledges the strength of baseball interest within the province's French Canadian community, and the often contradictory attitude of the church.

Members of the local church were not often sympathetic to new games. Stranded to an extent in British North America following the end of France's colonial rule in the 18th century, priests often stifled the natural spontaneity of their parishioners fearing that new enthusiasms paved the way for submersion in the dominant English culture. As conservative leaders within the French Canadian community they were wary of new sports emerging from the urban industrial cultures of England and America. Sport thus assumed a less significant role within Québécois culture, and adoption of sports and physical education in the school curriculum was belated. Horace Miner's 1939 study of the traditional parish of St-Denis, 30 kilometres north of St-Hyacinthe, recorded the general failure of sports such as baseball to gain a toehold in the community. Only ice hockey was played nearby.

Everett Hughes' 1943 study of another French Canadian community, somewhat more affected by external influences than that studied by Miner, recorded an encounter between the local priest and the manager of a small theatre who claimed to have lost money through the priest's objection to dance halls and movies. 'Why, that man even preaches against baseball,' he said. 'I ask you, what harm is there going to a baseball game after dinner on Sunday and sitting there outdoors until supper time. But no, that fellow doesn't like it. It's funny how in some towns the priests are good fellows, but in others some fellow like this one preaches against everything.' Hughes suggested however that the priest's real objection was less to baseball than to commercial amusements on Sunday.

But some Catholic spiritual leaders and students eagerly promoted baseball within Quebec, and through them the game entered largely French-speaking communities like St-Hyacinthe. The village of St-Hyacinthe (45 kilometres east of Montreal) was known for its religious and educational institutions. Less than 4 per cent of its population was British and their influence was marginal. In 1876 a baseball game was among the principal events of St-Hyacinthe's Saint-Jean-Baptiste day. It was a high scoring affair indicative of the players' rudimentary skills. How did such an obvious American game become part of celebrations marking French Canada's national day? Was this one more example of sport's ability to transcend narrow ethnocultural biases or was it, more negatively, a symbol of the people's alienation from their true heritage and culture?

The truth is really quite simple. Baseball was undeniably American but there was no French-speaking family that did not have some connection with the United States. Between 1850 and 1900 a half million Quebeckers left the province for jobs in New England's textile and resource-based industries. These French Canadian émigrés to America retained an abiding affection for their old homeland and a desire to sustain that connection in their Americanized children. One strategy was to send their children back for higher education in a French-speaking environment.

Young people came to these schools with a number of new habits learned in New England; foremost among these was baseball. Historian Donald Guay quoting from Forget's *Histoire du Collège de l'Assomption* says that in 1860 'la balle au camp' was known to students of the college due to the presence of American seminary and classics students. The Collège de l'Assomption would continue to attract young American students.

Jean-Marc Paradis' *100 Ans de Baseball à Trois-Rivières* includes an appendix documenting the background of 'franco-americains' students at the seminary of St-Joseph in Trois-Rivières between 1866 and 1900. Those who played baseball at the school include Arthur Gelinas from Marlboro, Massachusetts who studied medicine between 1883 and 1891, Jacques Emery from Hancock, a business student between 1887 and 1889, Emilien Delage from Webster who studied to be a priest between 1892 and 1899, Alfred Lachance who studied medicine from 1892 to 1899, Joseph Massicotte from Webster in dentistry from 1892 to 1899, David Olivier from Lowell, Massachusetts from 1892 to 1901, Alphonse Paradis and Pierre Mandeville, both from Grosvernor Dale, Connecticut, between 1894 and 1897, Wilfrid Caisse from Lowell, Massachusetts who studied pharmacy from 1895 to 1900, Hervé Forand from Webster between 1895 and 1898, Donat Lussier from Manville who studied to be a priest between 1895 and 1904, Joseph Hael from Lowell, Massachusetts from 1896 to 1897, Hector Peloquin from Southbridge who studied dentistry from 1897 to 1900, Arthur Harnois from Lowell, Massachusetts, 1899 to 1900, Joseph Gauthier from Worcester, Massachusetts, a civil engineering student between 1899 and 1903.

Baseball's growth among French Canadians was slow but steady. From Trois-Rivières and St-Hyacinthe, it spread to communities having no college. Historian Guay notes that a study of francophone newspapers in Quebec reveals 16 baseball articles from 1876 to 1890, 28 for 1891 to 1895, 110 for 1896 to 1898, 103 in 1899, and more than 200 in 1900.

An additional though more difficult influence to analyse is that of French Canadian ballplayers who played major league baseball in these early years. Onesime Eugene Vadeboncoeur who was among the 28 Canadians in the majors in 1884 is listed as born in Louiseville, Quebec. The birthplace of Sam LaRoque who played in 1888 is St-Mathias; that of Fred Demarais who played in 1890 is listed simply as province of Quebec. Pete LePine who played in 1902 is from Montreal. Ed Wingo's name at birth in Ste-Anne de Bellevue was Edmond Armand La Rivière. He played one major league game in 1920. Gus Dugas who played four seasons beginning in 1930 was born in St-Jean DeMatha. When Stuart McLean was researching his book *Welcome Home* a few years ago he could find no record of Dugas in St-Jean DeMatha and a local priest speculated that his family was among the countless who left for New England when Dugas was a child. Wingo's name change is in keeping with the practice of many who moved to the United States in this period. As for the rest their baseball success can have only one explanation. The game was not well enough organized in Quebec in these players' youth for them to have developed major league skills. In all likelihood those of French Canadian background learned them in New England where their families probably moved shortly after their birth. The English of Quebec did not move to the United States in anywhere near the same numbers and so their names do not appear in 19th century major league rosters.

Quebec-born French Canadians raised in the United States shared a common development path with that of many 19th century and early 20th century big leaguers born in the United States to parents originally from Quebec. Win Mercer, alias Mercier, born in 1874 in West Virginia, played primarily for Washington from 1894 to 1902. Eugene Napoleon De Montreville, born in 1874 in Minnesota, played major league baseball between 1894 and 1904. Philip Geier alias Giguere, born in Washington, D.C., played between 1896 and 1904. Les German alias Germain, born in 1869 in Baltimore, played for the New York Giants among several teams between 1890 and 1897.

Candy Lachance born in Connecticut in 1870 played from 1893 to 1905, and Abel Lizotte born in 1870 in Maine played with Pittsburgh in 1896. Napoleon 'Larry' Lajoie, Leo Durocher, and Clem Labine are 20th century franco-american ballplayers with Quebec connections. All of the above players were role models for aspiring young French Canadian ballplayers. In some cases their exploits would have been the topic of personal letters sent back and forth between relatives of these players in Canada and the United States. At the very least they piqued the interest of the French language press in Quebec.

Some of these players returned at least temporarily to the province and helped spread the game. One such was Del Bissonette who played for the Brooklyn Dodgers from 1928 to 1933. Born to Québécois parents in 1899 in a Winthrop, Maine logging camp, Bissonette played semi-pro ball in the Gaspé, the Beauce, and along the north shore of the St Lawrence but was ignored by big league scouts who seldom ventured into the French-speaking hinterland. That changed however when he hit .393 for Cap Madeleine in 1922. He was signed by the Dodgers to begin his gradual climb through the minor leagues. Another who returned was Mary Elizabeth Murphy, born in Rhode Island in 1894 and one of six children of an Irish mill worker and a French Canadian mother. Murphy led a barnstorming team throughout North America and for 50 cents one could see her pitch in Trois-Rivières in 1924 against a team representing St Maurice Paper.

Baseball in 19th century Montreal

The progress of baseball in Montreal is slower than in Ontario where it very early surpassed cricket and lacrosse as the leading summer game among working people, professionals and commercial classes. The latter two groups in Montreal were leaders in forming the Montreal Snow Shoe Club in 1840, the Olympic Athletic Club (organizers of the Montreal Olympic Games in 1844), the Montreal Lacrosse Club in 1856, and the Montreal Amateur Athletic Association (MAAA) in 1881 which incorporated among others the Montreal Bicycle Club (1878) and the Montreal Hockey Club (1875). Montrealers would also be prominent in establishing the national organizations that grew out of these local initiatives. A Montreal dentist, George Beers, was a key participant in the 1867 meeting in Kingston at which the National

Lacrosse Association was formed. Montrealers started the Canadian Football Association in 1873, the Canadian Wheelmen's Association in the early 1880s and an amateur hockey association for Canada in 1886. The MAAA was a leader in the 1884 formation of the Canadian Amateur Athletic Association, which ultimately led to the Canadian Olympic Association.

With so much going on it was hardly surprising that leading Montrealers either didn't have time for baseball or considered it too foreign to suit their national interests. Baseball to them was American, and Montreal's English entrepreneurs in particular were economic competitors with cities like Boston and New York for the developing western hinterlands.

Furthermore Quebec generally remained rural longer than Ontario (only 20 per cent of Quebec's population lived in cities in 1867), but because baseball's highest level of play was commercialized at an early stage of its development, urban areas were essential. While other sports were developing national organizations, baseball favoured the commercial benefits of a strong north-south orientation. This further limited Quebec's contact with the game's early hotbeds in Ontario. Despite the appearance of early baseball type games in the eastern townships in 1837 and the possibility of some residual memory of old Gallic bat and ball games these games appear to have no influence on baseball's later development.

Organized baseball in Montreal as in Halifax and other Maritime centres seems to have occurred only sporadically until the late 1860s. In fact in 1865, Montreal's municipal council had assessed a $5 fine on those playing cricket or baseball in local parks, possibly out of concern for pedestrians who might be hit by the ball. One of the few teams that did exist was but a subsidiary of a local football team. An important influence was undoubtedly American press reports which made their way into Montreal's English papers particularly after the formation of the all professional Cincinnati Red Stockings in 1869. During that summer a Montreal team scored 111 runs in a July match, a true indicator of the players' primitive skills. On August 12, a notice appeared in the Montreal *Gazette* addressed to that city's three teams by a well-known Canadian living in New York. He challenged Montreal teams to unite and come to his city for some exhibition games.

A week later the *Gazette* announced a match between an unnamed Montreal club and the Crescent Club of St John's, also of Montreal, on the

Montreal Baseball Club, 1871. Photograph Collection, Public Archives of Nova Scotia

Ball game at a picnic of the Typographers Union on Montreal's Sainte-Helene Island in 1874. Canadian Illustrated News, Public Archives of Canada C 61336

city's Lacrosse Grounds. The paper said 'this is the first opportunity that has been offered to many of our citizens to witness a first class game and we expect to see the field crowded to judge from experience whether it equals the more rough and tumble sport of lacrosse.' The paper said the game was gradually becoming popular among young men and its report of 23 August said 'both clubs are as yet quite young and many of the players have only had a few days practice. The spectators were very few. Play was as a rule, poor.' Notably, among the 18 players are four with French surnames, which appear to be Lameux, Goudreau, Lessard, and Labelle. Irish and English names make up the rest though some of these might also be French Canadian there being significant intermarriage at the time between French and Irish Catholics.

In Montreal baseball's growth was at best gradual in comparison to its popularity in Ontario. The 1876–77 *City Directory* lists the officers of the Montreal Association of Base Ball Players as John Davey of the Caledonia Club, A. Felix of the Red Stockings, J.B. Pardellian of the Excelsiors, Sam Ritchie of the Caledonians, and J. Mooney of the Union team. Historian Alan Metcalfe notes that in Montreal between 1867 and 1885 there were 46 baseball clubs in contrast to 63 snowshoe and 78 lacrosse teams. Baseball was not an insignificant sport but its teams had much shorter lifespans. Only 13 clubs lasted more than two years and none for the entire two decades. This may be due at least partly to the lack of national competition. Montrealers had shown a predilection for such championships in their other sports. The game's working-class popularity was pictured in a *Canadian Illustrated News* drawing of a ball game at a picnic of The Typographers Union on Ile Sainte-Helene in 1874. By the turn of the century children of working people had their own juvenile team, the Beavers, managed by labourers and a train driver.

One of the earliest city leagues in 1886 included the Montreal baseball club affiliated with the MAAA.

Five years ago anyone mentioning the word baseball in that city [Montreal] would have been arrested as a lunatic or looked upon as a man who was speaking a dead language. Today everyone in the City of Trouble is crazy about the American game.

report in a Buffalo newspaper in 1901 about baseball in Montreal

The game was slowly gaining entry into the mainstream of popular local sports. A year later a league member, the Gordons team, left the league and briefly competed in tournaments and went on tour as a professional squad. Professionalism and its attendant commercialism was necessary if Montreal was to enter the ranks of competitive baseball outside Quebec. But pro sports found powerful forces aligned against them in Montreal where many of the country's major amateur organizations got their start. This was partly a matter of drawing social boundaries. Natives who once competed in open snowshoe events were effectively segregated into professional events which lacked the prestige of amateur competition. Professionalism in lacrosse was the preserve of working-class teams like the Shamrocks—'the horny handed sons of toil' from the working class Ste-Anne's Ward.

Baseball lacked a solid English-speaking middle class élite to support it and so French Canadian promoters and enterprising English Canadians turned to those who could provide the necessary commercial support. One of these was Joe Page, a former Indianapolis ballplayer and friend of the great Canadian baseball star Tip O'Neill. Page came to Montreal in the 1880s to do promotional work for the Canadian Pacific Railway. The CPR understood the value of attracting sports teams. The New Westminster Salmonbellies lacrosse team had travelled by train to challenge for national titles. Within the settled areas of eastern Canada and the northeastern United States the train made it possible to organize intercity leagues in every sport. Such success could be realized in Quebec, Page reasoned.

Montreal had a long way to go to catch up with Ontario in baseball. Guelph, London, Hamilton, and Toronto all had teams (and often several of them) in organized baseball before Montreal. The city had a short-lived franchise in the International League in 1890. There was another attempt in 1895 but real involvement did not commence until 1897. 'A big fire in the Rochester ballpark put the club out of business,' Page recalled.

Bill Rowe and I had a club that had been playing exhibition games in Atwater Park. We asked the Eastern League if we could replace Rochester and they gave us the okay. All the good players were in the big leagues in the States and that's where the money was. If you had any money to spend, you hired a guy who had played in the big leagues that the fans had heard about and made him your player manager. We did that with Charlie Dooley, and we won the pennant in 1898.

Montreal's minor league team was part of organized baseball and so was made up of Americans with little or no connection to the region. Another professional league at the turn of the century in Quebec appealed more directly to French Canadian pride and not incidentally the desire to see French names in club executives and on the field. Historian Guay notes that 'la ligue provinciale' had as its executive F. Payette of the Mascotte club of Montreal, J.E. Sevigny of Valleyfield, and J.P. Plamondon of St-Hyacinthe. Farnham, Sorel, and St-Jean were the other league members. Of the 54 identified league players 39 were French Canadian.

The dynamics of Quebec baseball in the 20th century would reflect the life of the province as French Canadians sought a role in the sport's organizations and team line-ups.

Baseball in 20th century Montreal

When it was suggested that the Montreal Expos were the team of the 1980s a golden age of baseball achievement was promised but sadly never realized. Likewise a reputation as a team having the best minor league farm system in the 1990s creates expectations that may also be frustrated. Causes are often laid at the door of the team's supporters rather those who perform on the field. In this analysis Montreal is said to be too small a market to support a modern baseball franchise in the era of free agents, and the victim of a weak Canadian dollar that requires attendances of three to four million to be able to compete on a level playing field with American teams. At the

Baseball club from Hull, Quebec. In Robert Fontaine's book The Happy Time *the young protagonist asks, 'Is it true Hull is wicked?' His father replies, 'Not in the part where they play baseball.' Public Archives of Canada, C 80406*

root of this discussion is the idea that French Canadians have only passing interest in baseball and their nationalist goals have caused them to abandon a game too heavily loaded with English-speaking connections.

Yet baseball has been an integral part of French Canadian life for over a century. The golden age of baseball in Quebec and Montreal, the 1930s through 1950s, had two things lacking in the modern game in the province—championship teams and convivial, friendly ballparks which were community centres. It was also a period in which the Montreal Royals dominated the highest rung of the minor leagues as a member of the International League.

French and English Canadians alike turned out in huge numbers to cheer on the Royals, the leading farm club of the Brooklyn Dodgers, an organization overflowing with Branch Rickey's prospects. Between 1941 and 1958 the team finished first seven times, won the International League playoffs seven times, and won the Junior World Series competition against the American Association champion three times. In 1948 they drew 477,638 spectators whose social composition was certainly more refined than a 1901 Montreal baseball audience characterized by a Buffalo reporter as a 'frog eating bunch of flannel mouths . . . [with] French sentences chewed off in the middle, Irish "come-all-ye's", and English sputters'. Yet in one sense the nature of the crowd hadn't changed. During the Second World War it included young Jewish kids like Mordecai Richler who would recall years later that while his parents feared Hitler and his panzers, Richler and his buddies were wary that Branch Rickey and his scouts would spirit Montreal's best players away to the big club. Likewise Montreal's French Canadian population were the team's biggest boosters so much so that even though Royals games were only carried on an English-speaking radio station, at the end of each inning the announcer deferred to a reporter who provided brief updates in French. At the conclusion the reporter would turn the microphone back to the

> If my father had his choice, he would have preferred me to be an all-star player rather than to be here as an invited guest.
>
> Pierre Trudeau at the 1982 All-Star game in Montreal, referring to his father Charles, part owner of the Triple A Montreal Royals in the 1930s (Charles Trudeau died of pneumonia contracted at the Royals' spring training site in Florida).

game broadcaster with the comment 'Okay, Bill' and in time the radio voice of Montreal Royals broadcasts became known to thousands of French Canadians as 'O.K. Bill'.

Montreal's earliest baseball organizers were largely English-speaking. One of the first was the American Joe Page, who with Bill Rowe and Tip O'Neill, the Woodstock-born star of the St Louis Browns a decade before, operated Montreal's Eastern League franchise. In 1908 they sold out to Sam Lichtenhein (also owner of the Jubilee ice rink and, in a few years, the Montreal Wanderers hockey club) who hired future New York Yankee executive Ed Barrow to manage his team. Former National League hitting star Kitty Bransfield took over as manager in 1913. He in turn was followed by Dan Howley in 1915.

The franchise was sold after the 1917 season and interest reverted to lower-classified minor leagues which seemed to understand better than the owners of Montreal's team that French Canadians wanted a greater sense of identity with their teams. This was forthcoming in 1928 when a group headed by Athanase David and Ernest Savard bought the Syracuse International League franchise and moved it to Montreal. A new stadium, Delorimier Downs at the corner of Ontario Street, was the Royals' home. Over 22,000 fans including Judge Kenesaw Mountain Landis, the commissioner of baseball, were at the team's May home opener. The owners had the bad fortune of contending with the stock market crash and resulting Depression and by 1931 the club was in serious debt. Three Montreal businessmen, Hector Racine, Colonel Romeo Gauvreau, and Charles Trudeau bought the team and retained Ernest Savard as an associate. Frank 'Shag' Shaughnessy, an ex-American athlete and resident of Montreal since 1911 when he was coach of the McGill football team, was hired as general manager. Shag installed lights and, borrowing from the example of the National Hockey League, convinced the International League to institute a four-team playoff system as a way to generate additional revenues. Attending the team's

Frank 'Shag' Shaughnessy, Montreal Royals' executive in the 1930s and President of the International League 1936-60.

spring training site, Charles Trudeau contracted pneumonia and died shortly thereafter, a traumatic event that was to shape the life of his son Pierre Elliott. Montreal won the 1935 pennant but a year later as the team fell out of playoff contention Shaughnessy averted the fans' cries to fire him by accepting the presidency of the International League, a position he held for 24 years. He moved the league's offices from New York to Montreal.

At the end of the 1938 season the team was sold to the Brooklyn Dodgers and became their number one farm club. Normally such a sale to outsiders would alienate local fans, but the Dodgers wisely retained local management in the person of Hector Racine and attempted when possible to include French Canadians in the Royals' line-up. Most significantly they provided high calibre ballplayers and a winning tradition which was to last two decades. Montreal lost the Junior World Series in 1941 but fans were satisfied with the team's progress. In 1945 they won the league pennant with three French

Canadians playing key roles. Stan Breard was a slick fielding shortstop, Roland Gladu lead the league in hits (204) and runs scored (126), and pitcher Jean Pierre Roy lead the league with 25 wins and 139 strikeouts. The Royals lost the International League playoffs in seven games to Newark.

And then along came Jackie Robinson

In October, Hector Racine stunned the baseball world with the announcement of Jackie Robinson's signing to a Royals contract. Robinson would be the first acknowledged black to play in the modern era of organized baseball, though it's certain that despite baseball's unwritten apartheid rules, other blacks who disguised their identity as either Indian or Spanish Cuban had played before Robinson. Robinson's signing was really the work of Branch Rickey who understood the risk he was taking. If Robinson failed it might set the cause of integration back several years. On the other hand if Robinson excelled but struck out at his persecutors, reason might be found to continue baseball's race restrictions. But Rickey also knew that if Robinson succeeded both on the field and in his response to abuse, than the Dodgers could corner the market on much of the great black baseball talent then playing in their own leagues.

Rickey had visited Montreal many times over the previous years and caught a scent of fairmindedness and empathy among its citizens, many of whom felt ill treated by the English-speaking minority. There were no guarantees of acceptance but Rickey thought the gamble was worth taking, particularly since he knew there were few places in the United States that would accept a black player. Ironically, the Montrealers thought Racine had called the press conference to announce the transfer of the Philadelphia National League team to Montreal and so the news of Robinson's signing was not treated as an earth-shattering event. Once in Montreal, Jackie Robinson and his wife were able to rent the first apartment they visited, and throughout the season they were treated as a normal couple by their French Canadian neighbours. This more than anything else provided a security and normalcy that allowed Robinson to cope with the abuse he received in American cities. More than once general manager Mel Jones was visited by a shaken Robinson who would say 'Nobody knows what I'm going through.'

Just before the start of the 1946 season, the nervous International League president Shaughnessy warned Rickey about possible race troubles in Baltimore should Robinson appear and asked him to reconsider his decision. Rickey chose to ignore him, and Robinson debuted in Jersey City before a sell-out crowd of 25,000. He went four for five and electrified the crowd in the fifth with a bunt single, followed by a stolen base, and then went to third on an infield out. As the Jersey pitcher went into his motion, Robinson streaked for home but stopped halfway causing the pitcher to halt his delivery. The umpire called a balk and waved Robinson home. Afterwards he was mobbed by enthusiastic blacks, a scene repeated in other International League cities including Baltimore. The Toronto *Globe and Mail* for instance noted that 'Robinson's appearance on the diamond last night brought out quite a few of Toronto's colored colony.' So taken was Montreal with their new star that there was talk of signing Robinson to play football with the city's Alouettes that fall. Rickey turned thumbs down on that idea.

Robinson led the league with a .349 batting average, scored a league-leading 113 runs, and swiped 40 bases. The Royals defeated Newark in the first round of playoffs and then eliminated Syracuse to advance to the Junior World Series title against Louisville. Fans in the southern city taunted Robinson unmercifully and after winning the first game the Royals fell behind two game to one. Returning home the team was heartened by their fans' response. 'We discovered,' Robinson later said, 'that the Canadians were up in arms over the way I had been treated. Greeting us warmly, they let us know how they felt. . . . All through that first game [at home] they booed every time a Louisville player came out of the dugout. I didn't approve of this kind of retaliation, but I felt a jubilant sense of gratitude for the way Canadians expressed their feelings.' Nor was the fans' reaction the only uniquely Canadian touch. It was early October and the temperature dropped below the freezing point. Despite having taken a 4–0 lead the Kentucky team succumbed to the taunts of 15,000 fans and the weather, as Louisville's pitcher walked the bases loaded in the ninth and the Royals scored two to send the game into extra innings where Jackie

> ## Go back to Canada, black boy.
>
> Taunt directed against Jackie Robinson by Louisville fans attending the Little World Series against the Montreal Royals in 1946

Robinson drove in the winning run in the bottom of the tenth.

Milder weather brought out nearly 18,000 fans for game five and again Robinson starred as his booming triple in the seventh inning put him in position to score the winning run, but for good measure he bunted home Al Campanis with two out in the eighth. In game six Robinson was brilliant both in the field and on the basepaths and in the eighth the Louisville pitcher went through a routine of fake throws and real ones in an unsuccessful attempt to pick Robinson off first and second. The crowd howled. The Royals protected an early 2–0 lead for a series victory. Robinson was mobbed by jubilant Montrealers who cried 'Il a gagné ses épaulettes' and he was carried about the field in a scene so moving that sportswriter Sam Martin wrote, 'It was probably the only day in history that a black man ran from a white mob with love instead of lynching on its mind.' It remains the greatest moment in Montreal baseball history. Robinson went on to a Hall of Fame career with the Brooklyn Dodgers beginning in 1947 but the historic import of Robinson's role and play that season lifted a minor league title into the realm of the mythic. A solitary human being and a city confronted society's worst demons and for a brief moment slew them.

The golden years of Montreal baseball

Baseball in Montreal was even more popular over the next decade. In 1947, 442,485 fans watched the Royals and another great black star, Roy Campanella. Those numbers were topped in the minors only by cities in the Pacific Coast League which had a longer season. Montreal surpassed its attendance record the next year and remained almost constant in 1949. The 1948 team had Duke Snider for part of the season and their regular line-up included pitcher Jack Banta who won 19 games, Don Newcombe who was deprived of successive seven inning no-hitters by a scratch infield hit on 20 August, and the future television star Chuck Connors at first base. They went on to win the Junior World Series. Montreal finished third in 1949, a year highlighted by Sam Jethroe's league-leading 89 stolen bases, 154 runs scored, and 207 hits. The 1950 team

Montreal Royals player slides into third in game against Ottawa Athletics on 20 July 1952 at Montreal's Delorimier Stadium.
Public Archives of Canada and the Montreal Gazette, PA 151879

Montreal Mayor Camilien Houde throws out the first ball to open the 1953 International League baseball season in Montreal.
Public Archives of Canada and the Montreal Gazette, PA 151881

Montreal Royals 1955 International League pennant winning team. Line-up includes future Hall of Famer Don Drysdale, and Los Angeles Dodgers manager Tom Lasorda. Canadian Baseball Hall of Fame

featured Walter Alston as manager, and the 1951 team included Junior Gilliam, pitcher Joe Black, and George 'Shotgun' Shuba. Another Dodger legend, southpaw Tom Lasorda, joined the Royals team that won the 1952 pennant but lost in the playoffs to Rochester. The 1953 team won the Junior World Series with a roster that included power hitting Rocky Nelson, outfielder Sandy Amoros whose dazzling catch would save the 1955 World Series for the Brooklyn Dodgers, and Tommy Lasorda and his 17 pitching wins. The Royals' 1955 pennant-winning team had future Hall of Famer Don Drysdale for part of the season. Ironically Drysdale died in Montreal in 1993 while in the city to broadcast a game. Attendance remained near 400,000 until 1952 when it dipped to 313,000, and then began a descent which corresponded to the wider decline in minor league baseball interest in the 1950s as spectators stayed home enthralled with television and the parental responsibility of raising the baby boom generation. But at its peak the Royals paid for the Dodgers' entire scouting operation. Besides those listed above, other future Dodger stars who played in Montreal included Dan Bankhead, Don Hoak, Ralph Branca, and Johnny Podres.

When the Dodgers left Brooklyn for Los Angeles after the 1957 season Montreal's days were numbered. The 1958 Montreal team was strong enough to win the 1958 pennant but in 1959 the Dodgers' better players were sent to their new number-one farm team in Spokane. Just over 100,000 watched them in 1960 and in the same season that saw the International League lose Havana it also said goodbye to Montreal.

The Quebec Provincial League

Several rungs down the minor league ladder was the Quebec Provincial League, officially recognized by organized baseball in 1940 but a pioneer in baseball matters for many seasons before. The League had its roots in short-lived organizations like La Ligue Provinciale in the late 19th century. Baseball's growth in Quebec was restricted by poor spring weather and the realities of a largely rural society in which young men were busy in the hayfields in June and July. Baseball, says historian Merritt Clifton, was a late summer game played at tournaments at county fairs.

In the decade before the First World War the

Eastern Townships Industrial League once again attempted to organize matters, but teams were free enterprise creatures which sometimes ignored their formal schedules to play more lucrative exhibition games. Granby businessman, land speculator, and politician Homer Cabana ran one such team starting in 1909. Sixteen-year-old catcher Joseph Bousquet joined with Cabana's team in 1912 and spent most of his professional career in Quebec reinforcing local fans' identification with native-born stars. Bousquet fought in the First World War and afterwards played in St-Hyacinthe (1919), Farnham (1920), Drummondville (1921), Saranac Lake (1922), St-Hyacinthe (1923–27), Victoriaville (1928–30), Sherbrooke (1931–34) and three final years in the Provincial League, 1935–37. Another team of the era was the Dow Brewery Nine, 'none of whom,' says Clifton, 'ever brewed more beer than they drank'.

In 1920 an informal provincial league was formed with teams in Montreal, Granby, and Quebec City, their players drawn largely from local French and English talent. In 1922 the teams were accepted into organized baseball as the Class B Eastern Canada League under the presidency of Joe Page. League players included Ed Wingo from Ste-Anne de Bellevue, who started in a Montreal industrial league before a brief major league appearance, and Del Bissonette. Travel costs for a league strung out between Ottawa and Quebec City in 1923 were high, benefiting only Joe Page's CPR Railway. In 1924 seeking more profitable markets the league expanded to become the Quebec-Ontario-Vermont League but the two Vermont communities quit in mid season, as did the rest of the league at season end. The league had failed to tap the baseball-crazy Eastern Townships and was thus denied a central location where travelling teams could have played on long road trips. For the next decade baseball outside Montreal resorted to the informal circuit of challenge matches and tournament play.

The Provincial League's golden age began in 1935 and despite occasional interruptions lasted through 1969. It was organized by, among others, Sorel club owner Lucien Lachappelle, Judge Amadee Monette, Jean Barrette, Sherbrooke second baseman and secretary treasurer Amadee Roy, Granby's Homer Cabana, lawyer Patrick Delaney, and accountant R.A. Leduc. It was a more geographically compact league in which teams travelled by car and returned to their home town after a game. Its colourful membership in 1935 included not only Granby but also Lachine, Drummondville,

a Mohawk team from the Caughnawaga reservation, and two Montreal squads sponsored by Joseph Choquette and by the Montreal Police. The league relied not only on local ballplayers like Roland Gladu and Shorty Coderre but also American college players who, as amateurs, were not able to play in the professional minor leagues. The Provincial League was outside the structure of organized baseball but that didn't stop it from paying its players. It just had to be discreet with certain players. In 1937 the Sherbrooke team went on strike when paycheques were late and the players threatened to return to their US colleges for the football season. The hiring of American college players would be adopted in later years by semi-pro leagues from Edmonton to Nova Scotia. Another player source was hockey. Players like Alfred Thurrier, Doug Harvey, Maurice Richard, and Toe Blake were among those who over the next decade used baseball as their summer exercise. Harvey played four seasons (1947–50) for Ottawa of the Border League and averaged .344 before opting for a career as an all-star defenceman with the Montreal Canadiens.

Most significantly, in the summer of 1935 Granby signed a black player, Alfred Wilson, away from Chappie Johnson's barnstorming Colored All-Stars. There were no restrictions on such signings by teams outside organized baseball though they were relatively rare. (Satchel Paige played for one such team in North Dakota at this time.) In midseason Granby's team along with the rest of the Provincial League was in financial distress and with the help of Wilson, league teams played a series of games with a number of black barnstorming teams and the white House of David. Symptomatic of a long history of uneasy relations which publicly surfaced during the Oka Crisis in 1990, the season came to a dubious conclusion when the members of the Montreal Police team jailed some of the best Caughnawaga native players prior to a big series.

The Provincial League's real interest was in returning to the ranks of organized baseball and to that end they had to jettison some of their more unorthodox and progressive policies. Native-based teams and black players disappeared and the league took on the appearance of a traditional minor league, developing the next generation of major leaguers. One of those was Paul Calvert from Montreal who compiled a cumulative 10–6 record for Sherbrooke in 1938 and 1939 but who was gone by the time the league received official minor league status in 1940. Calvert was something of an iconoclast in the tradition of Moe Berg, a legendary

Wilfredo Salas, Normand Dussault, and Adrian Zabala of the Provincial League Sherbrooke team. Centre fielder Dussault also played hockey in the Montreal Canadiens organization.

catcher between the wars who specialized in languages and served as an atomic spy for the United States in World War Two. Calvert was a college graduate who spoke three languages fluently and studied accounting in the off season. As a major leaguer he was nicknamed 'Cyclops' because of his thick lenses which added to his mystique. Having spent $1,500 to obtain his release from the International League Toronto Maple Leafs after the 1948 season, he signed on with the Washington Senators where he not only won five games in a row but also lost 14 in a row, the fourth longest losing streak in American League history.

Another notable league player of the late 1930s was one-armed Pete Gray who hit .283 for Trois-Rivières in 1938 and later played part of a major league season with the St Louis Browns in 1945. Possibly Quebec's most committed native son was Roland Gladu, whose major league career consisted of only 66 at-bats for the Boston Braves in 1944. Between 1928 and 1951 however he played in, among other locations, Granby (1928–31), Montreal of the International

Paul Calvert and Adrian Zabala of the Provincial League Sherbrooke team. Montreal-born Calvert later played in the major leagues.

League (1932–33 and 1945), Lachine of the Provincial League (1935), Quebec of the Canadian-American League (1936–37 and 1941–42) and with Quebec of the Provincial League (1938–40), St-Hyacinthe of the Provincial League (1947), and Sherbrooke of the Provincial League (1948–51).

The league received organized baseball's blessing in 1940 but in the process was required to drop certain players like Howie Moss of Granby who had played outlaw baseball in North Carolina in 1937. Outlaw leagues unlike independent ones were those that defied the authority of major league baseball by signing players banned for one reason or another from organized baseball. In the process all the league's players regardless of their transgressions were declared ineligible to play for a team in organized baseball. With Canada at war, the 1940 season was a murky one. Both Drummondville and Sherbrooke folded before season end and the remaining teams joined the Canadian-American League in 1941. Among the rookies of 1940, however, was Jean-Pierre Roy. Of French Canadian and Scottish ancestry, he continued a tradition of Québécois baseball stars who would achieve at least a modicum of major league success, but who more than anything else demonstrated the popularity of baseball in the French language community in this era. Roy enjoyed his baseball years in Quebec and won 25 games for Montreal's International League team in 1945, but he was less than enamoured by his stay with the Brooklyn Dodgers in 1946. He fled to an outlaw league in Mexico, found the conditions unacceptable and returned to talk things over with Branch Rickey who had threatened to banish him from organized baseball. Rickey was sympathetic to Roy's desire to play in Montreal rather than Brooklyn, and so the young Quebecker was allowed to become a part of the Royals' 1946 Junior World Series champions.

During the war the supply of imported ballplayers dried up and with it the quality of baseball generally. In Quebec, French Canadians took a dim view of conscription and participation in the war effort which they concluded was at least partially an English Canadian excuse to die for the British Empire. As a result there was less pressure to enlist and this freed many like Stan Breard and Normand Dussault, both later prominent in the Provincial League's great years of the late 40s and early 50s, to play baseball. Veterans like Del Bissonette and Alfred Thurrier were able to extend their careers. The old Eastern Townships League was revived as the Ligue Commerciale, along with the Yamaska or Richelieu League, and eventually the Provincial League again in 1944. Author John Craig who was at the time serving in the Canadian Navy recalled that the military was furious that so many apparently able-bodied French Canadians chose the charms of baseball rather than service in Europe. At one point, he says, they rushed a number of good ball-playing serviceman to St-Hyacinthe in an attempt to demonstrate the superiority of military men. According to historian Merritt Clifton there is little evidence that such aggressive enlistment helped St-Hyacinthe's cause.

The Provincial League's post-war golden age

Baseball enjoyed unparalleled success following the war. There were several factors. Stars like Ted Williams and Joe DiMaggio returned from war service, and with Jackie Robinson's arrival in Brooklyn the doors opened to an entire generation of black players. Fans, desperate for frivolous entertainment following six years of war, flocked to major and minor league parks throughout North America. In 1946 Quebec-based teams played in the International, Canadian-American, and Border leagues, but not all of these were financially viable. Granby and Sherbrooke of the Border League found transportation costs prohibitive.

More advantageously for owners the limited number of major league jobs in relation to the incredible proliferation of teams and players allowed them to drive down the wages of established players. In Mexico, Jorge Pasquel, the wealthy owner of a baseball league and a man with links to the leading presidential candidate, concluded that the best way to get his man elected was to inflate the calibre of local ball. His associates signed American major league regulars like Sal Maglie, Max Lanier, Mickey Owen, Adrian Zabala and Danny Gardella together with lesser lights like Roland Gladu and Jean-Pierre Roy. When Pasquel's man, whose campaign emphasized his role in bringing big leaguers to Mexico, became President the baseball project fizzled. The former major leaguers were stranded. They had been suspended from organized baseball for jumping their teams. Law suits were filed and one by Gardella might have brought about unlimited free agency had he not accepted an out-of-court settlement. In the meantime some players formed barnstorming squads while others considered offers from leagues willing to defy organized baseball.

The outlaw spirit was best exemplified by the suddenly revitalized Provincial League. Roland Gladu returned from Mexico in the latter part of the 1947 season and convinced entrepreneurs like Maurice Guillet of Sherbrooke and Albert Molini of Drummondville that investing in banned players would pay off in better baseball and larger gates. Guillet signed Gladu, Zabala and three Cubans Gladu had met in Mexico. Molini then signed Lanier and Gardella. Black baseball stars also flocked to Quebec, encouraged by the response to Robinson and aware that few American minor league teams would give them a chance even though baseball's doors had been opened to their race. Also signing on were gifted French Canadian stars like Stan and Roger Breard, and Jean-Pierre Roy who had been released by the Montreal Royals.

In 1949 Drummondville added Sal Maglie along with ex-big leaguer pitcher Tex Shirley, Negro League star Quincy Trouppe, and a future major league power hitter, 18-year-old Vic Power. Significantly 5,000 asbestos industry miners in the region had gone on strike on 14 February and for four months Quebec was racked by social conflict that touched the provincial government, the Roman Catholic Church, the mine's anglophone managers and American owners, and the until then quiescent union movement. It changed forever the cosy relationship between these powerful interests and is considered a significant turning point in the evolution of French Canadian nationalism. It's likely that local interests anxious to deflect attention from the strike put money into the baseball team. This money was needed in mid season. With Gardella's case working its way through the judicial system, major league owners suddenly offered to reinstate the Mexican League rebels. Lanier and Zabala were among the many who abandoned the Provincial League. In Drummondville Molini, confident that he could earn expenses back through gambling, responded by giving Maglie $15,000 to finish the season. Farnham emerged as an unlikely challenger and met Drummondville in the playoff finals. Even here the league's outlaw status was confirmed as many Farnham players were suspicious that the favoured Drummondville team went at less than full speed to ensure a final game in their own park. Such questionable behaviour was reinforced by reports

that $20,000 was wagered on the final game won by Drummondville behind Sal Maglie.

Maglie returned to the New York Giants in 1950 and resumed his reputation as a pitcher willing to throw inside, hence his nickname, the 'Barber'. Alfred Molini resigned as president of the outlaw Provincial League and organized baseball welcomed the rest of the League back into its fold by granting them Class C status for 1950.

Semi-professional baseball outside the structure of organized baseball continued in Quebec. Scotty Gibbs' baseball experience was typical:

> There was a big fire in Rimouski in 1950 where a third of the town burnt. Rimouski called me to pitch for them the following year. It was late in summer. I went up and pitched the last four innings on a Sunday night against Causapscal. I pitched to 12 batters as no one got safe on first. I got paid $70 for pitching the four innings. They offered me $70 a week, room and meals paid in a hotel, plus ten cents a mile for travelling. Causapscal had eight American players and their catcher had been given a try-out with the Brooklyn Dodgers. The weather was bad up there so I never went back.

Except for a two-year hiatus the Provincial League survived until 1969. Its days in organized baseball lasted until 1955. At a time when teams like the Boston Red Sox still had not integrated their line-up the Provincial League was aggressive in pursuing black stars. Vic Power hit .341 and .334 in his two Drummondville seasons. He was signed by the New York Yankees to integrate their team. It never happened. His flashy 'hot dog' style in the outfield earned him few friends among baseball's conservative establishment and he was bypassed in centre field for another phenom, Mickey Mantle.

Meanwhile Branch Rickey had moved over to the Pittsburgh Pirates in 1950 and a year later he appointed Sam Bankhead as the first black manager in the history of organized baseball with the Pirates' Provincial League affiliate in Farnham. Former Negro National League players were signed including Josh Gibson Jr, son of the immortal black catcher. The league's early start in hiring black players began to fade by the mid 1950s, and as the calibre of play declined the league hired locals to generate fan interest. St-Jean won the final pennant of the league's organized baseball phase in 1955; for two years thereafter attempts to re-organize the league on a semi-pro basis failed.

From 1958 to the end of the sixties the league at times relied exclusively on local talent and in other seasons hired ex-pros whose salaries stretched the league's meagre gate receipts to the breaking point. Among those who played at the league's end was former big leaguer Tim Harkness, with Granby in 1969. That season saw the debut of the Expos. Unfortunately for the Provincial League that team, and its stadium, Jarry Park, were within easy driving distance of the very Quebec fans who had supported the Provincial League. A new era in baseball had dawned at the same time as one that began in the late 19th century died.

Montreal Expos: good, bad and the ugly

By today's standards Montreal would never have been awarded a franchise. Local interest was lukewarm.

Only two French language reporters from *La Presse* attended the 1968 expansion gathering. There were no English media. Finances were suspect. Among the six investors only industrialist Jean-Louis Levesque and distillery magnate Charles Bronfman had any real money. Most significantly the team didn't even have a ballpark. A deal to use the Autostade had collapsed. Finally, contenders Buffalo and San Diego seemed to have the necessary votes.

Montreal succeeded because the decision was made at a time when baseball's stock was at an all-time low. Owners wanted expansion dollars but the $10 million they were asking seemed excessive to many prospective cities. (Eight years later when the American League accepted Toronto the franchise fee was only $7 million.) A year before, baseball's esteemed *Sporting News* had promoted both Montreal's and Toronto's cause. And it's possible the two could have gone into the majors together except that Toronto didn't even bother to apply. In

Montreal Expos President John McHale, former Prime Minister Lester B. Pearson, Expo player Rusty Staub, and Expo owner Charles Bronfman, at the Expos' spring training camp in West Palm Beach, Florida, March 1970. Public Archives of Canada, PA 127426

Montreal's favour was the recent success of Expo '67, Canada's major Centennial event. City councillor Gerald Snyder helped line up backers, with the behind-scenes support of the Los Angeles Dodgers who had warm memories of Montreal as the Dodgers' chief farm club in the forties and fifties and its support in the integration of organized baseball. Sentiment usually plays little part in baseball's decision-making but this was perhaps the ideal time to combine that emotion with a city's willingness to invest the necessary dollars. Perhaps at no other time in baseball history could Montreal have landed such a franchise, and it allowed the majors to promote themselves as the first major American-based professional league to make the jump to the international stage. A quarter century later, the National Basketball Association and the National Football League would still have no international playing representation.

The final piece to the Montreal baseball puzzle was the recommendation of Russ Taylor and Marcel Desjardins that the team consider Jarry Park for its home games. On the verge of losing the franchise, they drew up blueprints showing that the 3000-seat park could be expanded to 30,000. National League president Warren Giles inspected the site while a local amateur game was in progress and gave his direct approval, touched perhaps by the response of some in attendance who saluted him as 'Le gran patron'.

Unlike their Canadian cousins, the Blue Jays, the Expos have gone through all three stages of an expansion team's evolution. The first stage, which generally lasts five to ten years, is the growing pains of a line-up crowded with major league veterans and untried rookies. The second stage is the maturing of these rookies and new recruits into a competitive line-up. In the case of the Blue Jays this stage has characterized the team since 1983. The final stage which represents the end of any claim to expansion status is the decline from the team's first success. It is a punishing stage marked by small crowds and ballplayers actively avoiding the team in contracts restricting trades to that city. Free agents show equivalent disdain. When this organization returns to

some measure of success it does so with the mark of an established team.

The Expos' first stage lasted roughly from 1969 through 1978. Their first few years were a love-in with the city. Playing in tiny Jarry Park, a host of temporary heroes inspired fanatical support. Bill Stoneman hurled a no-hitter in the team's ninth game ever, before a sparse crowd at Philadelphia's Connie Mack Stadium. 'Le grand orange'—red-haired Rusty Staub—hit thirty home runs in 1970 and is recalled fondly by Quebec fans to this day. Ron Hunt accumulated a record number of trips to first base by refusing, when at bat, to duck away from balls thrown in his direction. Claude Raymond, the team's first French Canadian star, was fourth in the National League with 23 saves in 1970. Fans danced in the aisles of Jarry Park, visiting Torontonians marvelled at their ability to sit at a Canadian sporting event with a beer in their hands, and the Expo uniforms which appeared drearily conservative by the 1990s were, in those first few years, the sartorial talk of the League. To top it off the team competed well into late summer of 1973 for the National League East pennant. On one memorable weekend that summer they beat Fergie Jenkins and his Chicago Cubs when Fergie was at his peak. The sell-out crowd bellowed its approval. By 1976 however Jarry Park's attractions had worn thin as the floundering team drew just over 600,000 and Montrealers flocked instead to the Olympic Games.

The new Olympic Stadium was seen as Montreal's salvation and its occupancy roughly corresponded to the second stage of the team's evolution, as a young team battling through the 1979 season before losing the pennant on the last weekend of the season to the eventual World Series champion Pittsburgh Pirates. Through the 1994 season the Expos have never won as many games, 95, as they did that year but despite fewer wins in the two seasons ahead they emerged as one of the dominant teams in baseball. Pitcher Steve Rogers, catcher Gary Carter, and fielders Andre Dawson and Tim Raines were the foundation of what many said would be the team of the eighties. In 1980 they lost the pennant, again on the last weekend of the

> It's tough to be a Prime Minister in a country when you know Gary Carter could be elected tomorrow.
>
> Pierre Trudeau at the 1982 All-Star game in Montreal

Gary Carter of the Montreal Expos. Following the 1980 season Carter said his goal was to bring a championship to Canada, but his only World Series ring would be earned as a member of the 1986 New York Mets. Montreal Expos

The Expos valiantly remained competitive until 1991 when their .441 won-lost average was the team's worst since 1976. At some point in those years the Expos had entered into an expansion team's third and final stage. The descent was completed on 13 September 1991 when a 50-ton chunk of concrete from a support beam fell outside the stadium forcing the team to play the remainder of their games on the road—not necessarily a bad thing because attendance had never returned to its early eighties heights of over two million. There was general speculation that the team would move if Charles Bronfman could not find local ownership. He was successful in that quest and it led to a rapidly improving Expos team led by a crop of talented young players including Larry Walker, Moises Alou, Marquis Grissom, and Delino DeShields. In 1992 the Expos had entered into a new phase of success, marking their transition from an expansion squad to an established one.

The Expos' future in Montreal nevertheless remains uncertain and perhaps always will be. No matter how successful their local recruiting cam-

season, and again it was to the eventual World Series champion Philadelphia Phillies. In 1981 Rick Monday's homer in the fifth and deciding game of the National League championship series, on what has since been dubbed Blue Monday, ended another season. Appropriately Monday was to play for the eventual World Series champion, Los Angeles Dodgers. The Expos completed the 1982 season with the best four-year record in the National League (331–261) but it was a familiar story as the 1982 World Series champion St Louis Cardinals won the National League East pennant over the Expos. By this time Gary Carter had become arguably the most popular athlete in Canada with endorsements surpassing even those of Wayne Gretzky. He even lived in Montreal in the off season. But the club with so much promise just barely broke the .500 level in 1983 and a year later went 78–83. The Expos began breaking up their team with Carter going to the New York Mets, Dawson opting for free agency after the 1986 season, and finally Tim Raines departing in a trade to the Chicago White Sox in 1990.

Steve Rogers of the Montreal Expos. Montreal Expos

paigns are they will continue to be a team dominated by English-speaking Americans and Spanish-speaking Latins. Even here the magic of pitcher Denis Boucher whose arrival attracted large crowds in 1993 had worn off in 1994; only 12,000 came out to see an early season start. For the largely nationalist French media the Expos will always remain outsiders even though this élitist perspective is not held by the average French-speaking sports fan who is generally a better baseball fan than the local corporate class. Many argue that the Expos' only hope is an outdoor ballpark of the Camden Yards variety in a part of the city near their most rabid French-speaking fans. Such a goal is probably a dream in a cash-starved environment in which Canadian dollars are at a distinct disadvantage to the American ones of their National League counterparts. Blaming Montreal's fans for the team's fate however disregards the history of baseball in Quebec and the certainty that a team capable of overcoming the curse of the Rick Monday home run would again capture popular interest.

Postscript:

The owner's family

At one time a virtual first family of baseball, the Griffiths of Washington and Minnesota saw their central role in baseball organization diminished with the sale of the Twins in the mid 1980s. The dynasty traces through baseball Hall of Famer Clark Griffith and the sad decline of a former minor league ballplayer, Jimmy Robertson. Various accounts suggest Robertson was either born in the Shetland Islands, the Cayman Islands, or on a ranch near Helena, Montana and that his neighbours were the parents of the actor Gary Cooper. Robertson was a career minor league ballplayer whose career ended in Montreal where he and his wife Jane stayed to raise their seven children. It was a difficult existence and cirrhosis of the liver ended Jimmy's life at the age of 42. Jane's sister had been somewhat more successful in her choice of partners, marrying Clark Griffith. Griffith was a former big league ballplayer who first played in the

majors in 1891, managed for twenty years, and owned the American League Washington Senators from 1920 to 1955. (Recent magazine stories can't agree on who married who, suggesting in one case that Jimmy Robertson had married Clark's sister, while another says Clark married Jimmy's sister.)

During Jimmy's illness Uncle Clark Griffith had in 1922 brought two of the Robertson children to Washington from Montreal to live with his family. These two children, Thelma and Calvin, were later adopted by Clark and assumed his surname. The rest of the Robertson children came to Washington in 1924 after their father Jimmy died. Like Old Testament characters all of the children were put to work building Clark's baseball lineage. The line was passed directly to Calvin (born in Montreal in 1911).

He was a Senators bat boy when the team won the 1924 World Series, and a catcher at George Washington University. He operated Senators' minor league teams in Chattanooga and Charlotte. He came to Washington in 1941 to run the concessions at Griffith Stadium, and assumed the club presidency on his uncle's death in 1955. He moved the team to Minnesota after the 1960 season, changing their name to the Twins. The American League responded by putting an expansion team in Washington which also eventually departed and is today the Texas Rangers. In Minnesota Calvin was overwhelmed in the 1970s by the challenge of free agency which drove up the price of players. He lost many of his stars to free agency and in an angry diatribe alienated many of his remaining players by claiming he had moved the team to Minnesota because only 15,000 blacks lived in Minneapolis and they were notoriously poor baseball fans. Calvin's son, Clark Griffith II, went into law and argued baseball's first arbitration case in 1974.

Calvin's sister Thelma was an equal 26 per cent partner with Calvin and team Vice-President and Assistant Treasurer. She had married the pitcher Joe Haynes and their son Bruce eventually became a Twins Vice President. After the Griffiths sold the team, he scouted. Calvin's oldest sister Mildred, who had remained in Montreal to help look after brothers

> With baseball every spring there is a sense of new beginning, that lovely feeling of pulling yourself together. God, when you hear that first game on the radio, you know summer is coming.
>
> Corinne Griffith Pillsbury, daughter of Calvin Griffith

Bruce, Sherry, Billy and Jimmy, later married Joe Cronin, a player, manager, and later an American League president (1959–73). Their son Tom was also involved in the Twins operation.

Of the Robertson boys, Sherry, born in Montreal in 1919, was the most successful, playing with the Senators from 1940 to 1952 before becoming their Assistant Farm Director. He was appointed Farm Director in 1956 and held that position until his death in 1970. While Minnesota's team name derives from the twin cities of Minneapolis and St Paul, it surely is not a coincidence that the name appealed to Calvin because his twin brothers Billy and Jimmy both began working for the team in their youth. Billy sold hotdogs in Washington during the 1933 World Series and eventually became Vice President of Stadium Operations. Jimmy was Vice President of Concessions and his son was the Twins' travelling secretary during the Griffiths' ownership.

The Griffiths' sale of the Twins prior to the 1985 season not only ended Calvin's direct participation in the game (he retained an ambiguous advisor role), it also resulted in the departure or firing of the rest of the Griffith-Robertson clan.

CHAPTER EIGHT

Baseball in One Canadian Community

Bowmanville's council chamber was filled on the evening of 11 June 1868 as the friends of bat and ball games met to re-organize their town's cricket and baseball teams. Frederick Cubitt was elected president of the cricket club, and William R. Climie vice-president. A committee was appointed to sod the bowling crease on the ground adjacent to a drill shed recently constructed in the north end of town (on present-day Carlisle Avenue) in response to the Fenian troubles. Two years before, a band of Irish sympathizers had landed at Hull's Marsh as part of a larger strategy to strike at British interests in North America.

Officers were then selected for the Live Oak base ball club which played on the grounds north of a factory on Thursdays and Saturdays at 6 p.m. Climie was appointed president, Benjamin Werry vice-president, and Samuel Burden secretary-treasurer. The popularity of the two bat and ball games in 19th century Bowmanville and their fates was a story repeated throughout the Dominion.

Bowmanville, 40 miles east of Toronto, was then a rural community of several thousand people. It had not only several mills to service the surrounding countryside but a brick industry which had earned it the sobriquet of 'the brick town'. Settled in the 1790s by late loyalists from the United States, it derived its name from a pioneering local merchant, Charles Bowman. The chief protagonists in this story are Cubitt and Climie, whose backgrounds, political allegiances, and sporting fancies tell us much about life in 19th-century Canada.

At first glance they appeared to share a common affection for games and good fellowship despite conflicting politics—proof that a person may disagree with his neighbour on the resolution of public issues but still be capable of civic fraternizing. They were teammates on Bowmanville's Number Three Curling Rink and represented their town against Orono, and in 1871 played together in matches leading to the awarding of the Caledonia Medal. They even united as members of the married men's team (dubbed invariably the Benedicts in reference to Shakespeare's confirmed bachelor who is cozened into marriage) in less serious matches with a bachelors' rink. As residents of a small town their family lives were intermingled. News accounts of the day provide scattered reference to their garden parties and personal milestones.

In the summer they were proponents of the two popular bat and ball games of the age. In the success and failure of these two games was the larger story of the evolution of Ontario society and its political beliefs, in which Mr Cubitt and Mr Climie represented competing ideals.

Frederick Cubitt: a born fighter

Frederick Cubitt, born in 1819, arrived in Bowmanville as a 13-year-old in the company of his two brothers and father, a medical doctor. They had left behind their estate of Erpingham in the English county of Norfolkshire and settled on 400 acres of land on which part of Bowmanville now stands. They built a saw mill four miles northeast of Bowmanville where Frederick's older brother Richard, in J.B.

Turn-of-the century baseball scenes from Bowmanville, Ontario. Courtesy Bowmanville Museum.

Fairbairn's words, 'kept bachelors hall . . . and led a right jolly life' being 'a great favourite with his acquaintances'. The flirtatious and frivolous life of the village squire was tempered for Frederick however by the Upper Canada Rebellion of 1837.

He was shaped by his experiences as an 18-year-old militia man marching to Toronto to meet William Lyon Mackenzie's rebels. At this young age Frederick was cast on the path of dyed-in-the-wool conservatism. The rebellion marked a turning point in Canadian politics. The established Tory clique dubbed the Family Compact was eager to retain their political and land-owning privileges. Opposite them was a new business and small land-owning class interested in issues such as non-sectarian common schools, freedom of religion, an expanded vote and guarantees of tenure of office. Cubitt was on the side of old privilege but he was an active public servant who later served at different times as mayor, magistrate, school trustee, clerk of the Division Court, and Town Treasurer. His lifelong interest in the army was eventually rewarded with the position of Lieutenant Colonel of the 45th Battalion.

Cubitt touched upon national prominence at an election rally in Darlington in September 1867. John A. Macdonald wanted to be the last speaker of the day and Cubitt was called upon to delay proceedings. 'Mr F. Cubitts [sic] rose and attempted to speak against time,' the Globe reported. 'The uproar and calls for the Premier became so great that Mr Cubitts was unheard, but so long as the time of the meeting was frittered away Mr Cubitts' purpose and the Premier's was answered, and so when Mr Cubitts became tired of gesticulating, hat in hand he took a hearty laugh at the fun of the affair. In this way he wasted half an hour or more.' An exasperated George Brown representing the reform cause finally took the platform thus allowing Macdonald to go last. Cubitt was, in Fairbairn's words, 'a born fighter and most persistent in the attainment of any object he undertook. . . . He was a good conversationalist, delighted in reminiscences and told many a good story of the olden time. He was a great all round sport and patronized most kinds of athletic games. He was a keen curler and a good cricketer.'

William Climie: a liberal public spirit

On the reformers' side was William Roaf Climie, born in 1839 in Simcoe County and brought to Bowmanville five years later. His father, John, was a Congregationalist minister and raised in William a reforming temperament which included the values of liberal democracy and fury at alcohol's role in destroying family life. 'John Climie was,' says John Squair, 'a type of clergyman who mingled politics with religion in a way which has gone pretty well out of fashion in our day.' He purchased The Messenger newspaper in 1855 and renamed it The Canadian Statesman. It was defiantly Liberal in political outlook and this tradition was carried on by his son. As editor of Bowmanville's leading newspaper, William Climie was one of hundreds of delegates to the Great Reform Convention held in Toronto's Music Hall on 27 June 1867, three days before Canada's formal recognition as a nation. He considered the great Canadian Liberal of the 19th century, Edward Blake, to be an intimate friend.

'It is a duty men owe to themselves and their fellow men to encourage a liberal public spirit,' William Climie wrote. If not, 'A man may be born, grow up, pass through life, and die in a place, and yet that place never receives one particle of benefit from his existence. He might as well have never lived. A turnip or cabbage would exert just as favourable an influence on the public mind as he does.'

For his part Climie perceived the 1837 Rebellion in a different light from Cubitt. Though he regretted its violence and acknowledged that Mackenzie may have been an extreme man, nevertheless, Climie concluded, 'had he not taken the stand he did we might still be under the iron heel of the family compact.'

Cricket's days of glory in Darlington

The potential for conflict in their political beliefs would not appear to have much relation to the sporting arena but the fate of cricket and baseball was inevitably tied up in the changing nature of Ontario society. The Colonel's game of cricket had much in its early favour. It came into Canada and the United States in the 18th century. Records exist of matches in Manhattan in 1751 and the game had been introduced into virtually every part of settled Canada by the early 19th century. As early as 1825 an English immigrant, George Anthony Barber, was coaching students at Upper Canada College. They would become the leaders of Ontario society but their political allegiance was conservative and many were highly placed members of the Family Compact. They guarded their rights and control of public matters and not surprisingly in 1836 the Toronto *Patriot* newspaper noted that 'British feelings cannot flow into the breasts of our Canadian boys through a more delightful or untainted channel than that of British sports. A cricketer, as a matter of course, detests democracy and is staunch in his allegiance to his King.' The failed rebellion of 1837 was a victory for their class and the ascendancy of their bat and ball game was reflected in a level of organization that allowed cricket matches to be held between Toronto and New York in 1845. On 24 and 25 September 1849 the first ever official international cricket match held anywhere in the world took place between Canada and the United States in New York City.

Youth from around Ontario, educated at Upper Canada College and themselves members of the colony's conservative ruling class, brought the game back to their small towns where they in turn assumed leading positions as members of the judiciary, local government, the medical profession, etc. Cricket was their informal means of maintaining the social cohesion of a local ruling class as well as providing opportunities to meet with the élite of Upper Canada. On 8 June 1846 Charles Neville, Secretary of the Darlington Cricket Club (composed of players from Bowmanville and the surrounding Township of Darlington) invited the Cobourg Club to play a friendly match on the 17th in Bowmanville. In the absence of either rail or adequate roads in what was still a pioneering community, the Cobourg players travelled by stagecoach and the steamer *America*. Bowmanville's cricket ground was located in a park-like setting just north of the then settled part of town in an area bound by present Lowe Street, Beech Avenue, and Centre Street. Deer roamed the still wooded far end. In this idyllic setting Bowmanville won by 24 runs. A month later on 18 July as recorded in Hall and McCulloch's *60 Years of Canadian Cricket* (published in 1895), the visit was reciprocated. John Bailey, an unordained preacher of the Methodist Episcopal Church in Bowmanville, scored 10 and 36 in his two at-bats. It might have been more except that a heavy shower between innings caused Bailey to tumble head first on his wicket. The umpires could not decide whether or not he was out. Both sides agreed to await the verdict of the Toronto Cricket Club which later ruled against Bailey. Twenty subsequent runs were deducted though Darlington still won the match. Darlington's other notable batsman was Thomas C. Sutton. Sutton was truly representative of his class. He was a shareholder

in the Port Darlington Harbour Company, a church warden at St John's Anglican Church in Bowmanville (other parishioners were the Cubitts, whose family hound lay on the church floor beside their pew, and Charles Neville), a future Quarter Master of the Third Battalion of Durham, and a municipal councillor in 1853. Sutton was recognized as an outstanding underhand bowler and a stiff 'bat' who played on many select province teams against Upper Canada College.

The Toronto *Globe* recorded additional matches by the Darlington club in 1848, and Hall and McCulloch described a 16 September 1852 match in Toronto in which the bowling of Sutton and Cubitt was 'dead straight at the wicket. . . . in them Darlington had a pair of excellent bowlers.' Darlington's leading batters were Sutton and John H. Holmes, a lieutenant in Cubitt's and Sutton's Third Battalion. The line-up of their Toronto rivals included J.O. Heward, J. Helliwell, C.J. Rykert, Dexter, Madison, and George Barber all of whom had or would shortly have international playing experience. By the time of Darlington's match in Toronto on 11 July 1855, the Bowmanville-based team had been together for at least ten years, a considerable period of organizational success but not surprising given the élite membership of the team. They counted on public support of a type that politicians would extend to successful professionals in the 20th century. The lieutenant-governor Sir Francis Bond Head and Lady Head attended the 1855 match and the *Globe's* reporter hoped not only that the cricketers would prove 'the efficacy of physical training in assisting mental requirements', but that more women could now attend following Lady Head's example, 'for the modern cricketer puts forth his best energies by the consciousness that bright eyes look upon his efforts'.

The Darlington club retained its exclusive membership at the time of the 1855 match, which marked the first reference to St John Hutcheson, an Upper Canada graduate, splendid fellow, club secretary, and a prominent Bowmanville lawyer. The club's president Frederick Cubitt was into his second decade of play. These Darlington cricketers played intersquad games or matches against other towns whose line-ups conformed to their social class. They had little interest in proselytizing the game's merits to the general public which in Darlington consisted of about 8,000 people in the 115-square-mile township.

In 1857, despite good showings against Toronto and Whitby, the *Canadian Cricketer's Guide* reported that 'the strength of the Darlington club is necessarily weakened by the recent establishment of two other clubs in town—the "Franklin" and the "Union".' At this point the trail of bat and ball games in Bowmanville grows cold for a decade. This is a period of startling change in the fortunes of cricket and baseball. Despite its head start in popularity, cricket was coming under increasing popular competition from baseball.

Whatever momentum cricket had as a better organized and better played bat and ball game was forever forestalled by the American Civil War which among other things interrupted the incredibly popular and almost annual international series between Canada and the United States. By the time Americans were free to devote greater time to lighter amusements following the War, baseball had become the dominant game. The same factors that accounted for baseball's popularity and cricket's decline in the United States also applied to the majority of bat and ball players in Canada. Cricket's fate was compounded by the lack of interest shown by many cricketers, such as the members of the Darlington club, in expanding participation beyond their social class. Nevertheless it might have been expected that Canadians would show at least some regard for cricket's British connections and, particularly in this era of Fenian raids, exhibit a certain ambivalence to American symbols. Baseball, however, was at this stage still evolving away from its European folk roots. Its identity as an American game was not secure.

As for cricket its roots weren't deep enough, as in other parts of the British Empire they later would be, to fend off competition from other games. Cricket was being challenged by lacrosse which the Montreal dentist George Beers was promoting as Canada's national game. In 1867 Beers wrote, 'As cricket, wherever played by Britons, is a link to their home, so may lacrosse be to Canadians.' Beers would have liked to see lacrosse surpass baseball as well but its roots were too well established. In raising what had been the relative minor sport of lacrosse to a kind of patriotic prominence and expanding the number of summer team sport options, he contributed to cricket's decline.

Baseball comes to Bowmanville

Baseball's progress in Canada is equally instructive and again we return to Bowmanville. John Squair who was born in Bowmanville in 1850 says that 'so far as I

remember we did not, during my early youth, play base-ball or foot-ball.' Baseball's strongest centres in Canada in the 1850s were in the area west of Toronto. In eastern Ontario the game was spread not by fellow Ontarians but came north through New York State. Baseball was introduced into Cobourg by vacationing Americans who crossed Lake Ontario by boat from Rochester. From Cobourg the game spread throughout the County of Northumberland-Durham, the jurisdiction within which Bowmanville was located.

By 1865 the village of Newcastle, five miles east of Bowmanville, had a successful team, the Beavers. Players were drawn from Daniel Massey's manufacturing company. Led by a local educator, William Ware Tamblyn, they played other Ontario teams like the Ontario champion Woodstock Young Canadians, who scored 24 runs against Newcastle in one inning. Born in 1843 Tamblyn had attended the University of Toronto where he may have witnessed some of the first games in that city. Tamblyn was a pitcher but rules required him to deliver the ball from a point below the waist and to throw to a spot designated by the batter. To compensate for these disadvantages, Tamblyn stood erect in the pitcher's box with the back of his hand holding the ball towards the batter. As the arm came forward he turned it with a jerk thus giving the pitch speed and spin.

Some sense of baseball's more egalitarian character emerges in an 1868 story in Bowmanville's *Canadian Statesman* of ball-playing at a local picnic. In nearby Newcastle that year two nines of the local Beavers club under William Tamblyn and local storekeeper John Templeton played a match on the day celebrating the Queen's Birthday. There was no suggestion that cricket might be more appropriate. When it came to sports, cultural arguments mattered little.

Baseball had likely been played in Bowmanville for at least a few years before 1868, given the *Statesman's* reference to the apparent reformation of the 'old' Live Oaks. At a Dominion Day celebration they defeated a factory team from the Upper Canada Furniture Company 49–9. Later in the day that same factory club lost by only a run to a team of players from the surrounding Darlington Township. These results lend credence to the conclusion that the Live Oaks were an older and more polished squad. Of significance as well is the identification of the Darlington name with baseball rather than cricket. In August, the Bowmanville Sons of Temperance Nine including Ben Werry, William Climie, a local school teacher Malcolm McTavish, and members of the McMurtry and Burden families defeated a picked nine from town whose members included Tom Shaw. These individuals would not only be important in baseball's growth in the 1870s but represented a different set of values from those of the early generation of cricketers, being either small business men or middle class professionals.

In early October of 1868 the Beavers won a spirited series between Newcastle and Bowmanville. A year later a juvenile squad, the Victorias, was formed in Bowmanville. That same year a third major summer sport, lacrosse, was organized in town with the local team travelling to Rochester. In August Captain Bill's lacrosse club from the Six Nations 'dressed in costumes of varied and fantastic appearance and ornamented with feathers, paint and other decorations peculiar to the red men' visited Bowmanville.

In 1870, with time running out on cricket's prominent place in the summer sporting calendar, an attempt was made to re-establish the game in the town under the chairmanship of lawyer Robert Armour. Frederick Cubitt was club president and the team included two baseball players: Climie and Samuel Burden. Merchants were requested to let their clerks play during the week and hope was expressed that Bowmanville would 'regain its old position on the cricket rolls'. The team's social composition was notably broader than before and at mid season there was even a married men (which included Cubitt) versus bachelors game.

In 1872 the *Statesman* noted, 'There having been no cricket ground in Bowmanville for a number of years, the noble game is almost unknown here now.' Some new arrivals in town organized a series of matches with Whitby but by now cricket had become a kind of warm-up or practice session for a majority of players whose first game was baseball. These included David Fisher, Climie, Burden, Thomas Shaw, R. McConochie, and W.J. McMurtry. The best bat and ball skills were now found among baseball players. When cricketers played a baseball game against the baseball specialists they were allowed to have three more men on the field . . . despite the fact that a year before Climie batting third for the Live Oak Baseball team had noted in his newspaper that 'beautiful muffings give evidence of want of practice.' At a firemen's picnic at the drill shed grounds in Bowmanville in summer 1872, teams from Hamilton, Cobourg, Oshawa, Port Hope, Lindsay, Napanee, Port

Perry, and Belleville competed in a variety of competitions from baseball to bands. In mid August, Bowmanville's junior squad, the Victorias, played the second nine of the Toronto Dauntless club. The Victorias' train arrived in Toronto around 10 a.m. and after touring the town the players were fêted at the Caer Howell House where the match was played on the cricket ground. Samuel Burden from Bowmanville umpired the match which the visitors won 32–30. The result was telegraphed back to Bowmanville and a large crowd gathered at the train station to welcome the Victorias home.

Cricket, baseball and the political landscape of Ontario

Cricket's decline and baseball's rise matched the changing political circumstances of Ontario. Just as the narrow conservatism of Cubitt could not be translated into the new political reality of Ontario which had incorporated American-inspired ideas of democracy, so also was cricket unable to stem the sporting tide in favour of baseball. The respective fates of the two games was never explicitly stated in political terms but the correspondence of an anonymous writer to the *Statesman* came close to making the point. Known only as the 'Gossiper', he may have been Dr King, a graduate of Dublin College, recently selected by Cubitt to become principal of Bowmanville's united High and Common schools. Regarding his personal politics he said that his sympathies 'have always been more or less with the great Conservative party', and on sporting matters he noted in the *Statesman* of 12 September:

> In Toronto, the week's event has been the visit of the English cricketers; and a good many of our townsfolk took advantage of the cheap trip, to pay a visit to the Toronto cricket ground. Some folk, doubtless, were disappointed in the play; for it must be admitted that, with the exception of Mr Grace [foremost among the English team was the great 19th-century cricketer W.G. Grace who is considered to be the Babe Ruth of his sport], the scores made by the Englishmen were nothing extraordinary. . . . The Torontonians made a much better stand against them than either of their former opponents; but there are few clubs able to compete against such an eleven. For my part I hope that the result of their visit will be that an additional impetus will be given

to the game and its votaries in Canada, as in my opinion it is a far more noble, exciting, and scientific game than Base Ball, which seems to have so entirely usurped its place amongst Canadians.

The Gossiper and his friends were swimming against the tide both politically and in regard to their favourite game. When the liberal election victories were reported in late August, Climie noted, 'the band played in the streets, and several new brooms being borne aloft . . . indicat[ed] that Durham County was now swept clean.'

Games represented the dynamic forces at work in a rapidly changing society moving from a pioneer rural status to an urban industrial one. Labour was in upheaval. The ruling classes that had put down the Rebellion of 1837 witnessed an incremental erosion of their former authority.

Baseball and William Climie's leadership

On the baseball front in 1873 the best team in Canada and one of the top ten in North America, the Guelph Maple Leafs, visited Bowmanville. They won easily, 25–2, and the local paper called it the most correct display of ball play ever witnessed in the town. 'They knew their business and played well together. Evidence is given of what practice and judgement will do,' said Climie in the *Statesman*. Baseball with its absence of social custom and social ties matched the ethos of the emerging business class who respected hard work, serious effort, and skill more than entrenched privilege. Later that season spectators paid 15 cents, and a half holiday was declared, for the 2 p.m. Friday game against the St Lawrence of Kingston whose line-up included five American professionals including a raucous battery of Rafferty and Dygert. The *Statesman* said that they were 'not fit for the society of respectable people— they are not wanted anywhere a second time, with their obscenity and profanity. Hotel keepers along the route do not want a second call from them, and at the railway station here one third of the party beat the bus driver out of his fare.' Kingston players berated the umpire every time a call went against them but the quality of their play like that of the Maple Leafs was guaranteeing baseball its place as the leading sport in North America.

In the process the ability of small towns to com-pete with larger ones paying more money for better players gradually began to disappear. Before this happened it was still possible for villages like Newcastle to receive an offer to play the best team in baseball. The Boston Red Stockings (who by dint of historical evolution are today's Atlanta Braves) were on tour through the American Midwest in the summer of 1873 and their manager and future Hall of Famer Harry Wright telegraphed John Templeton of the Newcastle Beavers to see if he would be interested in a match. Templeton agreed to Wright's terms but Boston accepted a better offer from Toronto and never made it to Newcastle. This was a period of high civic excitement. One week the celebrated New Orleans Minstrels visited, and another saw the arrival of L.B. Lent's Travelling World's Fair with its Bohemian Glass Blowers, the New York Circus Brigade, and Zanga the mysterious snake magician.

Baseball had also become the leading game among those in the country. People in Enniskillen were furious when they learned that the Seventh Line

Hall of Famer Harry Wright who arranged to bring the Boston Red Stockings to the Bowmanville area until he got a better offer from Toronto. National Baseball Library, Cooperstown, New York

had imported an outside player. In Newcastle tragedy occurred when a sudden thunderstorm sent players and spectators to the drill shed for cover. Lightning struck a flag pole on the south end of the shed and sent a charge across a wire to the shed, killing a Mr Burley from Clarke Township and injuring three Bowmanville visitors including Sam Burden. It was one of the first times anywhere that a spectator had been killed at a ballgame.

With the establishment of the Royal Oaks team in 1874 under the presidency of William Climie, baseball was confirmed as the leading sport in Bowmanville. This semi-professional team allowed Bowmanville to enter the ranks of the best teams in Ontario and reminded the remaining cricket establishment in the community of their former position in Ontario sport. The Fly Aways of New York were among those invited to play the Royal Oaks.

The game's successful commercialization in Bowmanville benefited many local businesses. Thomas Shaw was not only a rather stout member of the team but proprietor of the Alma Hotel where the players would eat and drink after the game. Drinking was a leading recreation of the day. Temperance societies fought for restrictions on the burgeoning number of taverns in the region. Climie followed his father's example in running an ongoing series of stories on the results of drunkenness even as members of his baseball team patronized Shaw's tavern.

The high point of the Royal Oaks' 1874 season was the visit of the Guelph Maple Leafs on their way home from their semi-professional championship victory in Watertown, New York. Bowmanville lost 20–7 but the Guelph players said that they were the best Canadian team they had played that year. Confirming that judgement, the Royal Oaks beat Kingston 13–10 in a game called after six innings so that Bowmanville could catch a boat back to their home town. They were declared the champions of eastern Ontario.

Meanwhile locals heard about the unfriendly behaviour of the Young Ontarios club of Bowmanville towards their rural cousins, the Eckfords of Orono. 'We wish to complain of the ungentlemanly behaviour,' wrote an Orono resident. He told the *Statesman* that Bowmanville had sent an unpaid telegram in response to a paid one, had delayed the start of their game until dusk and then taken the field with two members of the Royal Oaks on their team, had stolen the Eckfords' game ball, and refused to

repay the previous favour of the Orono players who had treated Bowmanville to dinner on a visit to their village. At least the Eckfords won 43–27.

Bowmanville and the Ontario baseball scene

Port Hope fans eagerly offered $5 bets which were 'gobbled up lively' before their baseball team played the Royal Oaks. It symbolized a new spirit of professionalism and seriousness in baseball which corresponded to its successful commercialization. This point is crucial to an understanding the different fates of baseball and cricket. For men of Cubitt's old-time Conservative lineage, cricket was a gentleman's game removed from crass material considerations except insofar as it provided an opportunity for a social élite to mingle and conclude their business in private. Baseball's supporters were from a different social class of merchants and middle-class professionals. They eagerly exploited the potential profits associated with public support of sports.

By 1876 the best senior baseball in Ontario had divided into two levels. The top level teams from London, Guelph, Kingston, Hamilton and Toronto played in the new Canadian league, while the others continued to play in an informal circuit of arranged matches. Unlike cricket, which had failed to attract widespread public support, baseball was by now rooted in the idle tossing of a ball between fathers and sons, attendance at community games, and simple play at school and workplace picnics and holiday gatherings. Climie was part of a liberal trend which found favour with a newly enfranchised majority of the population. Baseball was better able to capture that spirit but even its growth would reach limits that the public would find distasteful. Port Hope's visit to Bowmanville for a game with the Royal Oaks was marred by ungentlemanly street behaviour by Port Hope youths, 'no doubt influenced by alcohol'.

In mid August 1876 the premier Canadian team, the London Tecumsehs, visited Bowmanville's Drill Shed grounds. Their visit was a kind of last hurrah for Bowmanville's struggle to compete in bat and ball games with larger centres. What had begun with the Darlington Cricket club was climaxing in the struggles of the Bowmanville Royal Oaks baseball team. Baseball like other sports was slowly segregating towns into divisions of equal strength and commercial ability. It was no doubt a fair solution to the problem of competition but it also served to reinforce the

Bowmanville's baseball grounds on Carlisle Avenue where the game was played from the 1870s until the site was developed for Goodyear workers' housing in the early 20th century

cultural and economic withdrawal of smaller towns from their once ferocious rivalry with places like Hamilton and Toronto. The Tecumsehs with two future National League stars, Fred Goldsmith and Joe Hornung, won the hard-fought game and stayed afterwards to socialize. Such camaraderie and fraternalism would also soon disappear from the game.

The 1877 season brought Bowmanville to the forefront of Ontario baseball competition. London and Guelph had joined the American-based International Association and the field was now open for teams competing for the Ontario championship. Local papers persisted in calling it the Canadian title though no other provinces were represented. The team met in the *Statesman* offices in late February and William Climie announced his intention to play a diminished role in the team's operation, although his brother George continued to be involved as scorekeeper. The team's pitcher was Jim Schofield who worked in the Dominion Organ Factory on Temperance Street. He was locally renowned as one of the early curveballers, but 'He gave in easily,' said one of his teammates, 'always complaining of a bruise or sore arm so he could not always be depended upon.' Edwin Coleman's diary described the

opening of the local baseball season between the Royal Oaks and Ben Werry's Old Swamp Angel team. He described Werry as 'the hero of a hundred baseball scraps'. The Royal Oaks defeated Toronto and Markham early in the season but the London Atlantics, led by a future major league pitcher and umpire, Bob Emslie, easily beat the Oaks 19–0. Bowmanville lost a contest in Woodstock after walking off the field to protest an umpire's call that Coleman could only take one base after a wild throw past first. Scandal soured the Hamilton Standards' visit in August when Tom Shaw overheard two Hamilton players urging their manager to umpire that day's game because 'This team [Bowmanville] is too heavy for us and we intend to squeal and you can help us.' Squealing was a big part of Hamilton's game and they eventually forfeited it. At season end the *New York Clipper* noted that Bowmanville had finished third in the Canadian championship.

Baseball's changing face and Climie's departure

William Climie sold the *Statesman* in 1878 to the James family and so ended his association with the

Bowmanville's later baseball grounds in the 1940s (two views)

Royal Oaks, who were by now in any case a commercial enterprise of which he had little interest. Under the management of Jim Schofield season tickets were sold and new players welcomed. Without Climie's active administrative support and the publicity afforded by the town's leading newspaper, the team lacked the ability to compete at a semi-professional level and soon disbanded. The success of teams like the Royal Oaks however had made possible the game's penetration into the everyday lives of citizens in settings as diverse as amateur competition and family picnics. Its integration into the working life of common people was a symbol of the kind of

society reformers like Climie celebrated. By the 1880s baseball teams in Bowmanville represented the furniture factory, the organ factory, the merchants on the north side of King Street, millers, clerks, and shoemakers. On the other hand this new, freer, urban lifestyle brought a seamier side of Bowmanville to the surface. A bawdy house was closed in June 1880. The *Statesman* reported that 'residents of Queen Street, east of Ontario have been annoyed by frequenters of a house kept by Ellen Downey, a tall, coarse featured, black haired woman of considerable physical development, age about 40 years.' Mr Loscombe who had himself been run out of town a few years before after having an affair with the family's maid, appeared on her behalf at the court house.

The professionals who had played for the Royal Oaks found success in larger centres. Jim Wilcox was a local boy who first played baseball in Bowmanville as a 15-year-old in 1868. A decade later he played catcher for the Oaks without benefit of glove, mask, or protector but only a piece of rubber protruding between the teeth to prevent them from being knocked down his throat. After the team folded he went to Harriston where he played for the professional Brown Stockings and alongside Bob Emslie. In the early 1880s Wilcox went to the Cass Club of Detroit, making $150 for the season, and he later played in the American south before retiring in Cincinnati.

As for cricket it had become a game with no remaining connections to the everyday life of Ontarians. For a few it served as a symbolic reminder of a British past and had a specialized function as the property of a social élite. In other parts of Ontario prominent citizens joined cricket clubs in their later years as a means of stating their social success. It was a feeble gesture at best and no doubt men like Colonel Cubitt were saddened by the decline of their once popular sport to the margins of society, even if those edges were occupied by some of its most powerful patrons.

Baseball in its many forms of t-ball, softball, hardball, and slow pitch is played today by enthusiasts ranging from young children of four or five years of age to men and women. Cricket on the other hand has almost completely disappeared from Bowanville. And the Colonel and the newsman lie buried within a few feet of each other in the Bowmanville Cemetery— all the better to disagree on political and sporting matters into eternity.

CHAPTER NINE

Black Baseball in Canada

Unable to secure passage to Europe in the turmoil at the end of World War Two, the leading black writer of his age, Richard Wright, instead spent two months in Quebec. Living anonymously on the Ile d'Orleans in 1945, says historian Robin Winks, 'he savoured the financial independence that the mounting sales of *Black Boy* were giving him and sought out "a way of living with the earth" rather than, as in New York, "against the earth".'

The notion of Quebec as a kind of island of calm refuge in a continent of racial division was a popular one at the time and was fed by French Canadians' own emerging sense of themselves as a people apart. Perhaps in the struggle of the black community French Canadians sensed their own marginalization. In any case their province became the focal point for one of the great moments not only in sports but in 20th-century life. Even as Richard Wright was celebrating the relative freedom his time in Quebec afforded him, Branch Rickey, general manager of the Brooklyn Dodgers had a plan that would give ample public recognition to Wright's private experience. Rickey had spent the summer scouting players in the Negro Leagues for the ostensible purpose of starting a new all-black league that would play in Brooklyn's Ebbets Fields when the Dodgers were on the road. Among the players that impressed

> Rickey felt he had
> the ideal spot in which to
> break in a Negro ballplayer,
> the Triple A farm
> in Montreal where there is
> no racial discrimination.
>
> Tom Meany

Rickey was a former Southern California football star and ex-soldier, Jackie Robinson. Robinson was good enough to step directly into the Dodgers' line-up but the perhaps overly cautious Rickey wanted to see how he would handle himself in a setting somewhat out of the major league spotlight.

Montreal appealed to Rickey at least partially because he sensed its more tolerant, somewhat European attitude towards blacks. In this opinion he was on steady ground. Through the 1930s and 40s sports teams in the province of Quebec had been pioneers in broadening the base of participation. An all-black line of Herbie and Ossie Carnegie and Manny MacIntyre played in the Quebec Hockey League in the 1940s, and the Montreal Alouettes integrated the Canadian Football League over the objections of Toronto and Ottawa. Nevertheless there were limits to Quebeckers' tolerance. Black players were told to leave the province if they fraternized with women in the white community.

The black experience in Canada

Early black settlement in Canada was limited to specific areas and was the result of varying factors, all of which had origins in slavery. In New France local records indicate that there were 1132 blacks in slav-

Jackie Robinson as a Montreal Royal. National Baseball Library, Cooperstown, New York

ery in 1759. Following the Revolutionary War, Loyalists brought with them an additional 2,000 black slaves. At the same time, 3,500 free blacks who had fought on Britain's side in that war settled in Nova Scotia and New Brunswick. This number was supplemented by 2,000 slaves who fled behind British lines during the War of 1812. In 1793 Upper Canada had begun the process of ending slavery; in 1834 the British Parliament abolished slavery in all British North American colonies. The existence of an 'underground railroad' for fugitive slaves brought an additional 30,000 blacks into Canada by the time of the American Civil War. Many of these settled in the Windsor and Chatham areas in southwestern Ontario while a few drifted into the Toronto-Hamilton-St Catharines region.

Black settlement has also been documented in British Columbia at the time of the Fraser River Gold Rush of the late 1850s and early 1860s. In the period 1909–11 at least 1,000 blacks from Oklahoma moved into regions of northern Alberta. With a few notable exceptions blacks tended to live apart from the white community and so become nearly invisible in the country's history. The reasons were hardly sur-

prising. In Alberta blacks chose isolated areas in order to avoid white antagonism and to obtain enough land for the establishment of viable communities. In the more urban areas of the Maritimes and Upper Canada, their entry into integrated neighbourhoods and jobs was limited by whites who pointed to the social turmoil which followed the freeing of the slaves in the United States. A petition of white Albertans to the Prime Minister prior to the First World War declared:

> It is a matter of common knowledge that it has been proved in the United States that negroes and whites cannot live in proximity without the occurrence of revolting lawlessness and the development of bitter race hatred, and that the most serious question facing the United States today is the negro problem. . . . There is no reason to believe that we have here [in Canada] a higher order of civilization, or that the introduction of the negro problem here would have different results.

Within their own communities, blacks attempted to develop a self-sufficient economic base. The community was not large enough to support either an independent farming lifestyle or a series of independent businesses and many blacks were forced to seek manual labour in nearby cities or join the growing ranks of black porters who worked for the various railroads. While attempting to remain outside the larger white community—which was never as openly racist or segregated as the United States nor as welcoming and integrated as Canadians would like to believe—blacks gradually assumed some of the patterns of Canadian culture.

In the matter of games blacks adopted those which lacked significant social restraints. Though from the earliest days formal amateur baseball organizations banned black participation the game's growth has always found a way to bypass existing organizational structures when they interfered with emerging interests. This is because baseball's roots, like those of commercial entertainment, have been in the commercial support of businesses and working peoples' organizations and not in élite clubs or schools.

Blacks adopted baseball in North America not because they had been co-opted by the white establishment but almost despite that heritage. For baseball offered something that the white community was not prepared to provide, an entry into the kind of

William Galloway of the Woodstock Bains amateur team, 1898. Oxford County Museum

The 1916 Zee-Nut series trading card of Jimmy Claxton of the Oakland Oaks. Born in British Columbia, Claxton was the first black player to appear on a baseball card. Courtesy William Weiss

ation that had not as yet infiltrated either working-class or slave communities.

In Ontario where informal bat and ball games date back to the 1830s and where formal teams appeared by the mid 1850s, an all-black team, the Goodwills of London, was sufficiently established by 1869 to play the all-black Rialtos of Detroit. Indicative of the contradictory nature of race relations in these days was the schedule of the all-white Guelph Maple Leafs of the mid 1870s. In 1874 they played the Ku Klux Klan club of Oneida, New York at a tournament in Watertown, New York, and a year later hosted the all-black barnstorming team, the St Louis Black Sox. Still there were no moves to integrate white Canadian teams and in this they were following the American example. American players in the London Tecumseh's 1878 International Association objected to the present of black players on rival integrated teams and in 1881 members of the Guelph Maple Leafs refused to play if an American black professional, Bud Fowler, joined their team. Fowler eventually played several games for a team in nearby Petrolia.

Though Americans playing in Canada objected to black players, what did white Canadian baseball players think of their black rivals? Our clues are limited but one is particularly telling. In 1887 'Tip' O'Neill, who had grown up in Woodstock, Ontario and became the greatest Canadian ballplayer of the 19th century, presented a petition to St Louis Browns club owner Chris Von Der Ahe, on behalf of himself and seven Browns players, objecting to an exhibition game with the all-black Cuban Giants. It read, 'We the undersigned, do not agree to play against negroes tomorrow. We will cheerfully play against white people at any time, and think by refusing to play, we are only doing what is right, taking everything into consideration, and the shape the team is in at present.'

Matters of race were coming to a head in the 1880s. In the United States the dislocations following the Civil War and the discomforting changes from a rural to an urban society caused the public to look for a scapegoat. Why they should find it in a generally impoverished community that had played little role in these changes is puzzling. Some northern whites had exploited recently enfranchised blacks in the American south and thus gained a temporary power base there. Southerners responded by forming organizations like the Ku Klux Klan not only to drive out these 'carpet-bagging whites' but to

everyday normal life experiences that the wider culture shared as matter of birthright. Within Canada, at least in Ontario and the Maritimes, baseball interest was locally generated rather than a heritage of previous American residence. Bat and ball games were at too rudimentary a stage in their development and still largely an extension of British culture to have been adopted by black loyalists after the Revolutionary War or War of 1812. Though these games were better and more extensively developed by the time fugitive slaves were swelling the underground railroad, they were still largely a middle-class recre-

re-impose second-class status on and therefore economic control over blacks. In the north where many blacks migrated following their release from slavery, they were seen as competitors for low-paying jobs, and resented. Similar resentment was felt against every new ethnic group in America. But blacks were just not new, or poor, but different by virtue of their skin colour and that could be changed by neither continuous residence nor wealth. Only intermarriage would do that, and sexual tension as much as any factor was at the root of anti-black racism.

The 1880s witnessed the first generation of sufficiently gifted blacks who could challenge whites for baseball employment. The media as often as not responded by ridiculing black players. In 1887 the *Hamilton Spectator* described Newark's George Stovey and Moses Walker as a 'coon battery' and suggested that the 'coloured population have a monopoly of the calcimine business.' Of Stovey the paper said, 'He is everlasting smoking cigars when he is off duty and looks as if he had just succeeded in colouring himself a trifle.' The *Spectator* later blamed a careless editor for the story and apologized to the city's black population.

George Stovey, one of the most significant 19th-century black players, is referred to by several sources as a 'light complexioned Canadian' though more likely, according to historian Jerry Malloy, he was a native of Williamsport, Pennsylvania who briefly played for an Ontario team in 1885. Little else is known about his origins. He was the greatest black pitcher of the century, winning 30 games for Jersey City in 1886 and combined with Fleetwood Walker in 1887 in Newark as the first black battery on an integrated team.

While the Irish, Germans, Italians, and those from the Dominican Republic all suffered ridicule and stereotypical analysis at the hands of the established order, only blacks were forcibly banned from the game's organized structure. A few black players participated in the major league American Association in 1884. As the decade progressed others joined a variety of minor league teams. At the same time 'Jim

> Fowler . . . is one of the best pitchers on the continent of America . . . He has forgotten more about baseball than the present team ever knew and he could teach them many points in the game.
>
> The *Guelph Herald*

Crow' legislation in the United States was reducing blacks to segregated second-class status and, though organized baseball never implemented specific rules banning black players, it effectively joined the spirit of the times in discouraging teams from hiring blacks. The chief minor league, the International League of which Toronto and Hamilton were members, had been a leader in signing black players. In late 1887 it decreed that no club 'should promulgate contracts with colored players'. Canadians were complicit in these actions. Torontonians for instance had taunted Buffalo's Frank Grant with cries of 'kill the nigger'.

Several lower ranked minor leagues continued to hire blacks until the end of the century and George Stovey was among those provided with temporary employment. According to historian L. Robert Davids, black Canadian Alex Ross played in the Northern Michigan and Michigan State League from 1887 to 1889. In the latter year one of his teammates was Bud Fowler. According to Davids, Ross was one of over 70 blacks who played organized minor league baseball in this period.

Blacks associated with segregated white teams after the 1880s were relegated to submissive roles. The Toronto team which appeared in Rochester wearing maroon caps and shirts, and breeches with gold belts, attributed their victory to a 'very small fat coloured boy who they acquired in Syracuse as their mascot'.

Blacks interested in baseball pursued a parallel existence. Separate status was irritating but beyond the control of the black community. There was little if any communication between outposts of black settlement in Canada and so no network of either competition or support was available. This is hardly surprising given the paucity of baseball contact between one region and another in the larger white community. For black athletes in Canada adrift in a baseball netherworld the only response was quiet resignation, or creation of a parallel black baseball structure.

In the Maritimes where baseball was experiencing tremendous growth in the late 19th century, a variety of all-black teams were formed. These included the Halifax based Eurekas who lost only one match in the 1890s. Other teams included the Truro Victorians, the Dartmouth Stanleys and Seasides, and the Independent Stars and North Ends of Halifax. Perhaps the most noteworthy were the Celestials of Fredericton, New Brunswick who barnstormed with the Victorias of Halifax in 1891. By the end of the century there was an annual Maritime championship for black clubs.

Individual success stories became a model for discouraged blacks. In British Columbia for instance blacks had arrived during the gold rush and often intermarried with local natives. The product of one such relationship was Jimmy Claxton, born in 1892 in Wellington, British Columbia of mixed Irish, English, Black, French, and Indian blood. He left Canada as a baby when his family moved to Tacoma, Washington. He eventually played a few games with the Oakland Oaks of the Pacific Coast League, becoming possibly the first person with black ancestry to play in organized baseball in the 20th century and certainly the first to be portrayed on a baseball card, released by Zee-Nut in 1916.

That same season Ollie Johnson of Oakville, Ontario played senior baseball in his home town. In search of better competition he joined a Buffalo-based all-black barnstorming team, the Cuban Giants. In later years before his death in 1977 he became a respected elder statesman of Oakville recreation and was awarded lifetime membership in the Ontario Baseball Association. Another black, Hamilton-born Bill 'Hippo' or 'Hipple' Galloway, played all sports with mixed teams as a youth. In 1899 as a member of the Woodstock hockey team, he and teammate Charlie Lightfoot became two of the first blacks in the Ontario Hockey Association. That same year he played 20 games with Woodstock's Canadian League team. In St Thomas he appeared nervous and the crowd taunted him. His .150 batting average probably brought about his release but it wasn't helped by the comments of rival players such as Hamilton's

> We have organized the only club of Professional Colored Ball Players in the Country and intend making an extended tour, knowing that a colored club will be a novelty, as well as a new enterprise. Now what we want to know, is baseball a success in Canada?
>
> St Louis Black Sox manager in a letter to Guelph owner George Sleeman in 1883

McCann who refused to play against him. Despite the urging of many Woodstock players to stay and play for the hockey team, he left the town and the country to play baseball for the Cuban Giants.

So thoroughly had baseball in Ontario been subsumed in the continentalist perspective of organized baseball that by the early 1920s amateur baseball authorities in Ontario depended on American direction in the matter of allowing local blacks to play in their championships. At least in Saint John, New Brunswick it was possible for an all-black team, the South End Royals, to play in the city's South End League. Their star player was pitcher Fred Diggs and they won the South End championship every year in the early twenties, most notably in 1922 when they won the intermediate championship of Saint John.

In Alberta, a thousand blacks from Oklahoma settled in several isolated areas between 1909 and 1911. Most notable was Amber Valley, originally known as Pine Creek, near Edmonton. Despite hardships ranging from insect invasion to premature frost, seventy-five of the original ninety-five black homesteaders remained and cleared enough land to receive their homestead patents. Unlike many other isolated northern Albertan communities this one survived both the First World War and the Depression and in the process developed a lively local culture. After 1915 the community held its several-day picnic which drew people from a fifty mile radius for, among other things, its baseball games. Amber Valley soon had its own team which toured northern Alberta. Indians and Métis were Amber Valley's closest neighbours and in return for teaching black residents such northern survival skills as woodcraft and tanning, blacks taught the Indians, with whom many intermarried, how to play baseball. The black community in Amber Valley survived until World War Two but then began an inevitable decline as young people sought better jobs in Edmonton or Calgary.

Baseball between the First and Second World

The Coloured Diamond Baseball Team of Halifax, Nova Scotia. Photography Collection, Public Archives of Nova Scotia

The Amber Valley Baseball Team, c. 1950. Glenbow Archives, NA 704-5

Former Negro League player Jimmy Wilkes played for the Brantford Red Sox in Ontario's Inter-County Major League in the 1950s. In this photograph he is a member of the famous black barnstorming team, The Indianapolis Clowns

'Shanty' Clifford, shown rounding the bases after a home run in the Dominican Republic winter league, was one of many black players who played in Canada after the collapse of the Negro Leagues

Wars was characterized by the tours of black barn-storming teams. Most of these were American-based and they relished the opportunity to travel without restriction on the Canadian Pacific Railway. Author John Craig claimed that he had played on one such team, Chappie Johnson's Colored All-Stars just after World War Two. Though Craig was white, Johnson supposedly asked him to use lamp black to help him fill out a depleted roster. Barnstorming was a precarious existence and faded once the better black players were allowed to enter the major leagues.

In their heyday two of the better Canadian-based black teams were the Chatham, Ontario All-Stars and the Black Panthers. The All-Stars won a provincial title in 1934. Despite the fact that hotels often refused to accommodate them they toured the province playing before large crowds until the start of World War Two. One of the Chatham players was Ferguson Jenkins, a third generation Canadian whose family had emigrated to Canada from Barbados. While playing baseball Jenkins met Delores Jackson whose grandparents had come to Canada through the underground railroad. They married and their only child Ferguson Arthur Jenkins was born in Chatham's St Joseph Hospital on 13 December 1943. Eventually Fergie Jenkins would marry Kathy Williams, a descendant of Josiah Henson, the model for Harriet Beecher Stowe's hero in *Uncle Tom's Cabin*.

With the collapse of the Negro Leagues follow-

Once a fervent opponent of baseball integration, Montreal Royals manager Clay Hopper, shown with Sam Jethroe on 12 September 1949, came to accept the changing times. Public Archives of Canada, C 66757

ing the integration of organized baseball in the forties, some of its stars began filtering into the wide variety of well-paying semi-pro leagues in Canada. Luther 'Shanty' Clifford, Wilmer Fields, and Jimmy Wilkes all played in Ontario's Inter-County Major Baseball League in the 1950s and early 60s. Wilkes began playing pro ball with the Newark Eagles in 1945. He first came to Brantford in 1952 as part of a barnstorming team, the Indianapolis Clowns for whom Hank Aaron once played. Over the winter Wilkes was invited to play for the local team and his professional playing career in Canada lasted another ten years. Clifford, a Brantford teammate, played for

Sandy Amoros and Chico Fernandez of the Montreal Royals, 1953. The Brooklyn Dodgers continued to sign black players while other teams procrastinated. Public Archives of Canada, C 66866

the Homestead Grays and Kansas City Monarchs in the late forties. Like Wilkes he appreciated the opportunity to put down roots and was the Inter-County League's batting champion in 1956.

Fields was the probably the best of the former Negro Leaguers. He was the ace of the Homestead Grays' staff for whom he first played in 1940; he led that team to the last Negro National League championship in 1948. The major leagues could never offer enough money to attract him but semi-pro organizations like the Inter-County League were often willing to pay a big buck for one star player like Fields. He won the Inter-County League's Most Valuable Player award in 1951 in his first year in Brantford. In 1952 Jack Kent Cooke paid him $14,000 to join Toronto's International League team. He returned to the Inter-County League hitting .379 for Brantford in 1954 and .425 for Oshawa in 1955 before returning to the United States in 1956.

One of the last tours of an all-black barnstorming team was witnessed by Saskatchewan baseball historian David Shury in August 1963. While taking tickets at a local match between North Battleford Beavers and the visiting Unity Cardinals, he beheld what in other circumstances might have been passed off as a vision.

> I looked up and saw two old cars pulling up the hill and into the parking lot adjacent to the ball park.

Out of the cars poured 11 or 12 black persons who looked like they could be ballplayers. As they headed over to the ball park gate I recognized the leader as Leroy 'Satchel' Paige. I recall hearing that Paige and his All-Stars had been in the province barnstorming earlier that summer but I didn't recall hearing of anything from them in the past couple of weeks and I just assumed they had left the province.

According to Shury the players were out of money and unable to pay for gas to get to their next stop in Kamloops, British Columbia. Saskatchewan had always been friendly to Paige. He had first played in a tournament in the southern part of the province in 1931, and again in 1935. In 1958 as a member of the Cuban All-Stars he had pitched against the Saskatoon Commodores. By the 1960s baseball's open road was closing down, and the destitute travellers were helped on their way by two hastily arranged exhibition ballgames with Unity and North Battleford. One of Paige's all-stars, Sherman Cottingham, would return and play for North Battleford in 1964 and '65.

'When I watched the Tuesday night exhibition game in Abbott Field [North Battleford] that August 1963 evening,' Shury said, 'little did I realize that I was probably watching the last game played in Saskatchewan by a touring group of black barnstormers.'

> A lost date in Canada
> meant a large gate
> gone forever.
>
> owner of barnstorming
> Indianapolis Clowns reflecting
> on a rain-out in Montreal in
> the early seventies

CHAPTER TEN

Owners, Organizers, and Players in Ontario Baseball

At the 1884 conference of Ontario teams, George Sleeman of Guelph warned that 'they must stamp out the present existing professional tramp baseball system.' It was a theme that Sleeman raised only when he could not afford the costs of stocking his own professional teams. Sleeman's role at the centre of baseball's progress in Ontario was gradually being superseded by the game's growth in larger places like Hamilton and Toronto. Nevertheless, Sleeman ably demonstrated that larger events and conditions in both baseball and the outside world are themselves subject to the actions of the imaginative individual. This truism recurs throughout Ontario's baseball history.

Two years after his ominous warning Sleeman had stocked a completely professional squad to face the bigger Ontario cities. Local papers encouraged him, noting that 'the excitement attendant upon the Riel Rebellion in the Northwest had just subsided and the next best tonic to inflate public interest was a bangup baseball match.' When Hamilton and Toronto ignored his challenges and joined the American-based International League, Sleeman kept his team intact and sent them on a series of barnstorming tours through Ontario, Michigan, and Pennsylvania.

Returning from one of those trips in mid June the team was astounded to find a band and more than 5,000 fans at the train station. The club's mascot, a dog of the Dandie Dinmont breed, was hoisted aloft by spectators. The players were greeted by cheers and tossed bouquets as their carriage took them to the Wellington Hotel for a reception. 'Thus ended the largest demonstration ever accorded a ball club in Canada,' said a local paper, and thus ended as well Guelph's and George Sleeman's place at the forefront of baseball's ascent to North American prominence.

Parkdale Baseball Club at Toronto's Exhibition Grounds, 1888. Public Archives of Canada, PA 60605

Toronto baseball: owners, managers, and organizers

Toronto baseball's rise to importance had been gradual but by the mid 1880s it was about to surpass its competitors in southwestern Ontario. According to historian Lou Cauz the better class of Toronto citizens had long dismissed baseball as 'just a sandlot sport played by undesirables'. Such an attitude was reflected in local bylaws. As late as the turn of the century a study of Toronto's social conditions stated:

> A child eleven years old appeared in the Police court charged with the offence of playing ball on Sumach Street. The ball, a small rubber affair, was produced in court and the boy when asked why he did not bring the bat also, explained that he had no bat, and was playing with the ball and a piece of stick when the policeman interrupted him. There was no question as to the guilt of the accused. Hugh Miller J.P., fined the boy $2 or ten days in gaol.

These attitudes changed slowly as prominent Toronto businessmen like E. Strachan Cox formed a joint stock company in May 1885 to operate a franchise in the Canadian Baseball League. Toronto would be a leading minor league baseball centre for much of the next 80 years, shaping local enthusiasm for the game and the way in which the public perceived it. Cox was the first of a series of entrepreneurs who influenced the baseball fortunes of Toronto. Three of the more prominent were Lol Solman, Ed Mack, and Jack Kent Cooke.

Toronto-born promoter Lol Solman was part owner and manager of both the Royal Alexandra Theatre and the Toronto Maple Leafs Baseball Club, which he purchased in 1897. Leafs schedules were invariably stuffed into theatre programs. He married Emily Hanlan, sister of the world champion sculler Ned Hanlan, and used that connection to expand his interest in the Toronto Island ferry service and the amusement parks at Sunnyside and Hanlan's Point where the Leafs played for many years. In 1925 he sold many of his holdings to raise $750,000 to build

Guelph baseball team and typical late 19th-century stands and spectators. Guelph Civic Museum

Unknown Toronto amateur team prior to World War One. Boy in the dark suit and tie was killed in the war, according to his niece Kate Hastings who retains the photo. Courtesy Kate Hastings

Ingersoll of the Western Ontario Baseball League, 1905

Barrie, Ontario amateur team, undated. Simcoe County Archives

Toronto Baseball Club, 1901, including the future New York Yankees general manager Ed Barrow. Public Archives of Canada, PA 52574

Picnicking at the Hanlan's Point ball fields in 1928. Toronto's International League team was by then playing in Maple Leaf Stadium at the foot of Bathurst Street. Hanlan's Point ballpark was the site of Babe Ruth's only minor league home run, 5 September 1914. Public Archives of Canada, PA 54473

Ike Boone at bat in Toronto's Maple Leaf Stadium, 1935. City of Toronto Archives, 36553

Toronto's Maple Leaf Stadium, 1933, home of the minor league Maple Leafs from 1926 to 1967. In the background is the Hanlan's Point stadium occupied by the club from 1910 to 1925. Toronto Harbour Commission Archives, PC 1/3/229

the 20,000-seat Maple Leaf Stadium on seven acres of land. Built in five months, it was ready for the 1926 season. The stadium was designed to accommodate an upper deck but several attempts to purchase existing major league teams failed and the extension was never built.

Another baseball entrepreneur of the era was Ed Mack, born in London, Ontario in 1856, who joined with 52 Toronto businessmen to buy the Toronto Maple Leafs back from the Toronto Ferry Company. He made a deal with the Toronto Street Railway Company to move the team to Diamond Park on Fraser Avenue near one of their routes in the King-Dufferin area. Mack, who made his money in the tailoring business, was at one time president of the Don Rowing Club. He was a prominent Mason who attended Glebe Road Presbyterian Church; Solman was Jewish. Normally in turn-of-the-century Toronto this would have consigned

them to different social circles. In baseball they shared a common enthusiasm.

In the 1950s prominent Toronto baseball entrepreneur Jack Kent Cooke's ownership of the International League Leafs reflected minor league baseball's greatest attendance and financial success. Owners like Cooke sold veteran players to major league clubs. Frustrated in his attempt to start the Continental League which would have rivalled the existing major leagues, he eventually sold the team and moved to the United States, where he became a media mogul and owner of several sports teams including the Los Angeles Kings hockey team and the Washington Redskins football team.

A second category of Canadian entrepreneurs were those who owned major league franchises. Possibly the most successful was Toronto-born Erastus Wiman who claimed that he had been the Toronto *Globe*'s first paper boy in 1846. By 1880 he

Opening game of Toronto's Eastern League season at Diamond Park at the southeast corner of Liberty Street and Fraser Avenue, 3 May 1907. Metropolitan Toronto Library, T 13354

was a wealthy New York railroad and real estate tycoon who retained an interest in Canada, using his money and telegraph service in the elections of 1887 and 1891 to back the Liberals in their quest for commercial union with the United States.

In the 1880s, following several unsuccessful efforts to promote lacrosse in the eastern United States, Wiman had purchased the New York Metropolitans of the American Association. Wiman used his wire service to fill Canadian newspapers with stories about the American professional baseball leagues, a process that continues to this day.

Other Canadian owners of major league teams have included the Quebec-born Joe Lannin, owner of the Red Sox around the time of the First World War, and Montreal-born Calvin Griffith, the adopted son of the Washington Senators owner, Clark Griffith.

A final category are those managers and organizers whose involvement was crucial for the game's growth. Baseball's commercial success in Toronto bred a generation of players with aspirations to enter the game's management ranks. Two of the most successful were George 'Knotty' Lee and Bill O'Hara. A product of Toronto's sandlots, O'Hara played for John McGraw's New York Giants before the First World War. McGraw called him one of the fastest outfielders in the game. During the war O'Hara served in Europe as an aviator with the Royal Naval Air Force. He later transferred to the infantry and served with the 24th Battalion, Canadian Expeditionary Force. At the battle of Ypres he suffered from shell shock and was invalided home. In mid season of 1927 he was appointed manager of the International League Toronto Maple Leafs. He immediately had the wall that extended from one dugout to another, in front of the grandstand, painted green in response to Toronto pitchers' complaints that this strip of concrete was too white. He was business manager of the Leafs

Letterhead advertising the Knotty Lee glove marketed by George 'Knotty' Lee. Canadian Baseball Hall of Fame

through the 1931 season and died while with the team in New Jersey.

'Knotty' Lee played an even more important role in 20th century Canadian baseball. Lee's life was devoted to the game. Born 12 May 1877 in Toronto, he grew up in an era in which lower class kids played baseball on the streets and upper class baseball fans like John Craig Eaton of the department store family travelled to Newark to watch Toronto's minor league team. Lee pitched a no-hitter for the Toronto Athletic Club in 1896 and two years later was playing for the Maple Leafs of the Eastern League. He developed a spitball throw which was said to twist or knot as it spun through the air, hence his nickname. He later played in Manchester, New Hampshire in the first decade of the 20th century. His managerial career between 1911 and 1940 included stops in the Canadian League with Hamilton, Toronto and Guelph, the Michigan-Ontario League with Brantford and Kitchener, the Ontario League with London, the semi-pro Central Ontario League with Kingston and Peterborough, the New England League, and the Can-Am League. He was a leading organizer of several of those leagues and in 1946 helped found the Border League. He was also the 1921–22 business manager of the Toronto Maple Leafs, a scout, and manufacturer of the Knotty Lee model baseball glove. He retired to Smiths Falls, a one-time Can-Am League town, where he ran the Lee Hotel, a gathering place of the sporting crowd.

Outside of the professional ranks were those Toronto baseball organizers who taught young peo-

'Wahoo Sam' Crawford, a future Hall of Famer, might have become a barber had not a Chatham, Ontario team offered him $65 a month to play baseball in 1899

ple the rudiments of the game. These coaches and teachers are largely the unsung heroes of the game. One example will suffice. Carmen Bush began his association as a volunteer with the Columbus Boys' Club of western Toronto in 1931. As athletic director he took teams to league, city and provincial championships. He retired in 1977 but continued to support baseball and was eventually recognized by his election to the Canadian Baseball Hall of Fame.

Countless similar champions of the game dot the Ontario landscape and from their generally unreported initiative came the Ontario baseball talent that has made the province the Canadian leader in supplying players to the major leagues.

Ontario players: the Assumption College connection

While most ballplayers got their training in publicly organized playground settings at least one school has played a leading role. Assumption College was founded in 1870 in Windsor, Ontario by Basilian priests as a high school level Catholic educational institution. The College records show that football and shinny were leading games in its first fall semester. Baseball was introduced in the spring of 1871, making Assumption one of the earliest schools in Canada in which the game was played. It was still a primitive game featuring underhand pitching, no gloves, and the batter's right to ask for certain pitches. In 1873 P. Murphy, a student from Akron, Ohio updated many of the game's features with his superior pitching skills. Prior to Murphy's arrival balls were tossed lamely to eager batters. Murphy put speed and finesse on his pitches.

Jay Clarke, John Upham, Reno Bertoia, Hank Biasatti, Pete Craig and Joe Siddall are six Canadian major leaguers with Assumption ties. Clarke was a descendant of United Empire Loyalists and the Wyandotte tribe of Malden Township near Amherstburg. Because his skin darkened noticeably in the summer he was, in the disparaging language of the day, nicknamed 'Nig'. An Assumption student in the 1890s, he moved on to minor league baseball where he is best known for hitting a record eight home runs in 1902 in a game won by his Corsicana, Texas team 51–3, though records of the game are considered suspect. His long major league career was interrupted by service in France with the famous Devil Dog contingent of the United States Marine

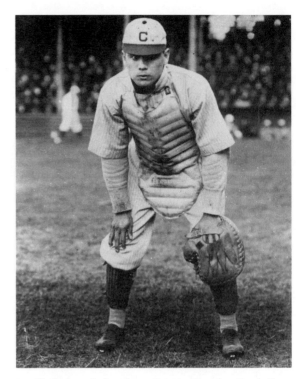

Jay Clarke from Amherstburg, Ontario. Canadian Baseball Hall of Fame

Hank Biasatti from Windsor, Ontario. Canadian Baseball Hall of Fame

Corps in World War One. After his baseball career he worked in the River Rouge Plant of the Ford Motor Company in Michigan.

Two other Assumption grads, Hank Biasatti and Reno Bertoia, were born in Italy. Their fathers had worked in Canada, returning to Italy to start families, and came back to Canada before the major post-war immigration from that country. The Italian community of the period was tight-knit in an era when non-English-speaking immigrants were looked on with suspicion and even fear. Schools like Assumption offered an opportunity to integrate into the larger community and still follow religious doctrines of home. Bertoia was a sensation in high school and 16 major league teams tried to sign him. He eventually accepted an offer from what were for all purposes the hometown Detroit Tigers. Bertoia debuted 22 September 1953 against Satchel Paige and the St Louis Browns in their last game ever in Tiger Stadium before they moved to Baltimore in the off season. Though he never became the superstar some predicted, he did find himself in an early season battle with Ted Williams for the 1957 batting crown. A story broke that Bertoia had used a mild tranquillizer to relax and overnight he was castigated as a virtual drug addict. He was later dealt to Washington. After baseball he became a successful school teacher.

> ... there was something about Bertoia, looking back on it, something that made you want things to happen for him and you felt maybe you wanted it more than he wanted it himself.
>
> Morley Callaghan

> ... one of the first native-born ball tossers ... Knight was a particularly handsome man with a black moustache and a physique that enabled him to bat and run and play the outfield.
>
> Frank Adams of Jonas Knight
> *London Free Press*, 18 October 1938

Some early Ontario players

The first successful Ontario ballplayers of the 19th century emerged from the generally informal network of sandlot baseball. One such was Joseph (Jonas) Knight who first played major league baseball with Philadelphia in 1881. He had earlier played for the Old Eries in Port Stanley, and then in Bay City, Michigan, London,

and Worcester. He was skilled at 'placement' hits and was excellent with the glove, recording a perfect fielding average with St Thomas of the Canadian League in 1899. When Fred Goldsmith, the hero of London Tecumsehs' 1877 International Association championship, returned to the city for an Old Boys' Reunion in 1937, he and London newspaperman Frank Adams visited Knight at his St Thomas home to reminisce about the early days of baseball in Ontario.

Another Ontario player may have played a part in one of America's literary masterpieces. According to a *Toronto Star* story from 9 October 1924 the original Casey who inspired Ernest Lawrence Thayer's poem 'Casey at the Bat' was born into a United Empire Loyalist family and raised in Newburgh, Ontario. Orrin Robinson (Bob) Casey played professional baseball in Minneapolis, Syracuse, and Detroit. He was renowned for his hitting prowess. Returning to Newburgh after the 1885 season he was invited to play for the home team against a Kingston, Ontario squad. The *British Whig* of October 15, 1885 described one of Casey's mighty blows which went foul and his failure to drive in the tying runs late in the game. Did Casey's reputation make it into a Boston paper where a relative was working? There is no evidence and Thayer always disavowed any connection to fact in the poem. But who knows?

For most of these early ballplayers baseball was their primary trade. As Ontario developed into an urban society, education played a more significant role in their upbringing. Harry O'Neill was born and raised in Ridgetown, Ontario and played baseball for the Ideal ball team. He graduated from the University of Toronto and found a teaching job in Alberta where he played on two consecutive provincial-champion teams in Medicine Hat. He was

Woodstock Actives team 1879. James Edward 'Tip' O'Neill (front row, second from left) was Canada's greatest 19th-century player. His fame as a member of the St Louis Browns in the 1880s was so great that future generations of O'Neills throughout the country often nicknamed one of their children 'Tip', as was the case for Thomas P. 'Tip' O'Neill, Jr, appointed speaker of the United States House of Representatives in 1977. Oxford County Museum

a member of the 1st Reserve Battalion championship team while serving in the Canadian forces in Europe during the First World War. Returning east to play for Windsor in 1920 he was signed by the Philadelphia Athletics. It was while pitching for the A's in 1922 that he apparently discovered the slider, a pitch that breaks low and away from the strike zone. At the time he had been trying to take the hop off his live fast ball so that his own teammates could hit the ball in batting practice. He later coached and managed in Salt Lake City and Boise, Idaho before retiring to Ridgetown. He recalled that the last big league batter he faced was Lefty Gomez.

Ontario baseball players in the thirties and forties

Though Jeff Heath was born in Fort William, Ontario of British parents on 1 April 1915, when he was one his family moved to Victoria, British Columbia and then on to Seattle, Washington while he was still a young boy. After high school in Seattle he was chosen the best amateur ballplayer in the Pacific Northwest and was among 18 players from all over the United States selected to tour Japan in the fall of 1935 for a pre-Olympic Games exhibition tour prior to their appearance in an exhibition game at the 1936 Berlin Olympics.

J. Geoffrey (Jeff) Heath

On the team's return to the United States, Heath was detained at customs when it was discovered that he was a British subject travelling on a British passport. For two days, until Lefty O'Doul intervened with the authorities, he was not allowed to leave the team's ship docked in San Francisco. Not long after Heath became an American citizen, cutting his final ties to Canada. After playing briefly for the Cleveland Indians in 1936 and 1937 he dramatically hit .343, with 21 home runs, and 112 RBI in 1938. Teammate Earl Averill warned him, 'Jeff, you're making a big mistake by hitting so well in your first big league season. You're setting a high standard for yourself, and the front office will expect you to do as well, if not better, next year and the following one too.'

Probably his greatest season was 1941 when he hit 24 home runs, 20 triples, and 32 doubles, the first American Leaguer to achieve this 20-plus cycle (and one not matched in that League until George Brett turned the trick in 1979). Heath was a member of the starting outfield with Joe DiMaggio and Ted Williams in the 1941 All-Star game. He was still going strong in 1948 when as a member of the Boston Braves, he hit .319, 20 home runs, and 76 RBI. Heath was destined to play against his old team, the

Indians, in the 1948 World Series until he broke his ankle at the end of the season.

A gifted hitter, Heath was a slugger off the diamond as well and his reckless behaviour probably kept him from an even better career statistical record. In 1939 he punched an abusive fan and threatened a Cleveland baseball writer. As a Washington Senator in 1946 he jogged down to second on an apparent double play that was fumbled by the infielder. He was immediately traded to the St Louis Browns.

After his playing career ended in 1949 he broadcast the Pacific Coast League games of the Seattle Rainiers. In 1956, furious at a number of technical glitches, he swore into an open microphone. Even though he apologized on air the station wanted to fire him but couldn't because he was hired by the sponsors. A KTVW station manager confronted Heath in the broadcast booth and the former ballplayer responded by throwing him down a flight of stairs to the grandstand level.

A year later Heath was accused of knocking down a Seattle construction worker in a café brawl. Another time he settled a matter out of court when a woman accused him of breaking into her house and slapping and shaking her. She told police 'she had formerly kept company with Mr Heath' but he ignored her desire that he remain away from her. Heath's raucous lifestyle ended shortly after his 60th birthday in Seattle with little indication that he ever gave much thought to the country of his birth.

Unlike Heath, four other major league Canadians of this era had strong Ontario roots. Dick Fowler learned his baseball in Toronto's Stanley Park playgrounds. After serving in World War Two with the 48th Highlanders of the Royal Canadian Infantry, he returned to throw a no-hitter against the St Louis Browns on 9 September 1945. In 1947 he struck out Joe DiMaggio four times in one game and in 1948 and 1949 he posted 15-win seasons. He retired to live in Oneonta, New York where he had played in the Canadian-American League in 1940, and died at the age of 50 working as a night desk man at the Oneonta Community Hotel.

Oscar Judd's baseball career began in Ingersoll in 1929. He moved up to senior baseball in Guelph between 1931–33, and spent the rest of the decade with minor league teams in Los Angeles, Columbus, Rochester, and Sacramento. He debuted with the Boston Red Sox in 1941 and remained there until traded to Philadelphia of the National League who were then known as the Blue Jays. He was with Philadelphia until 1948 when he was sent to their Triple A farm team, the Toronto Maple Leafs. After his baseball career he farmed and hunted jackrabbits in Ingersoll.

Phil Marchildon was the best Canadian pitcher of this era. His family history reflected the Canadian experience. Though most of the French Canadian émigrés from Quebec in the 19th century moved to New England, some went to Ontario and a large French-speaking community developed on the shores of Georgian Bay. Among the Québécois settlers who came to Penetang after the establishment of a naval base in 1817 were the Marchildons. The site of the nearby town of Port McNicoll was once known as Marchildon's Point. Several generations later, young Phillipe Marchildon who spoke not a word of French was pitching for his home town against small-town teams from Elmvale, Midland and Orillia. By the late 1930s he had joined Creighton Mines of the Nickel Belt League. Dan Howley, manager of the International League Toronto Maple Leafs, spotted him at a try-out camp in Barrie and he was a Leaf in 1939 and a 26-year-old rookie with the Philadelphia Athletics in 1940. Two years later he won 17 big league games before joining the Canadian armed forces. His plane was shot down while in action and he spent the rest of the war in a prisoner of war camp. Returning to the Athletics afterwards, he peaked in 1947 with 19 wins.

One of the more unlikely major league baseball heroes of the age was Gladys 'Terrie' Davis from Toronto who had played in the city's popular women's softball circuit of the 1930s. She joined the All-American Girls Professional Baseball League for its inaugural season in 1943 and was league batting champion. Davis was a combative player who once while at bat nearly came to blows with another Canadian, Bonnie Baker, the rival team's catcher. In later years Davis recalled her disdain of several unwanted sexual approaches by American teammates. 'I told one of them we don't do that sort of thing in Canada,' Davis said. She didn't stay long in the league, preferring the Toronto softball league.

Ontario ballplayers of more recent times

Ken MacKenzie was the New York Mets' only winning pitcher in their inaugural 1962 season, posting a 5–4 mark. MacKenzie had been raised in a sports

Phil Marchildon from Penetang, Ontario

family in the vacation community of Gore Bay on Manitoulin Island; Branch Rickey had a cottage on a nearby island. It was a sports-mad area. Men back from the war played in a local senior league which drew up to 1,500 fans in a town with a population of 700. MacKenzie's father was a good ballplayer and hockey player, and his uncle had played hockey with Aurel Joliet in Ottawa. In 1952 Ken enrolled at Yale University in New Haven and was among those college players who played in the Halifax and District Summer League. MacKenzie pitched for Truro.

Another Canadian ballplayer managed to overcome a serious challenge. Despite a heart attack after the 1970 season, John Hiller, a product of Toronto sandlots, returned to the Detroit Tigers in 1972 and a year later had his finest season with 38 major league saves (a record that stood until Dan Quisenberry saved 45 games for Kansas City in 1983). Hiller pitched 15 seasons for the Tigers in what was a transitional era for closers who often pitched several innings to end a game. He appeared in 545 games and saved 125 of them.

For a long period the Tigers' scouting network had a virtual monopoly on talent in southern Ontario and their signings included Reno Bertoia, George Korince, and Mike Kilkenny. Of them all Hiller was the most outstanding find.

Ferguson Jenkins

Ferguson Jenkins is one of just over 200 members of the National Baseball Hall of Fame in Cooperstown and the greatest Canadian ballplayer of all time. The righthander's 284 major league wins, 594 pitching starts, career strikeout-to-walk ratio of 3.2:1, and seven twenty-plus win seasons, are the stuff of record books and memories. He was the winner of Canada's leading sports award, the Lou Marsh Trophy, in 1974; Canadian male athlete of the year on four occasions; subject of a National Film Board documentary, *King of the Hill*, produced in the early 1970s; and recipient of the Order of Canada and entitled to use the letters CM after his name though he once confessed to writer Jay Teitel, 'To tell you the truth I haven't found out what it stands for yet.' The motto on the award is *Desiderantes meliorem patriam,* meaning 'they desire a better country'.

Jenkins grew up in Chatham, Ontario, a community of 40,000 just 50 miles northeast of Detroit. Prior to the Civil War it was a centre of anti-slavery activity and one of the northern termini for the Underground Railway which brought fugitive slaves to Canada. Among them were the Jackson family from whom Fergie is descended on his mother's side. It was from that side as well that he claimed two other traits. 'My father says I got my size [6 foot, 5 inches] from her and my precision pitching because she was so exacting,' Jenkins claimed in his

autobiography. From his father he inherited a love for baseball and, apparently, fishing which had been the family's occupation in Barbados. Ferguson senior was an outfielder for several Chatham teams and it was at these games that Fergie's parents met.

Even with its Underground heritage, in Chatham blacks were a minority. Perhaps this eased the process of integration. Jenkins has said he never experienced any racial anguish growing up in the community though he claimed he heard racial slurs when he played outside Chatham. In his youth he was a gifted

Fergie Jenkins holding the Cy Young Award he won as a member of the Chicago Cubs in 1971

boy soprano who credits the poise attained in musical competition with aiding his later success on the mound. His pin-point control he dates to pitching chunks of coal from Terry's Coal and Ice Yard near his house into the open doors of moving boxcars.

By the time Jenkins left Chatham in 1962 to begin his professional baseball career he was already something of a local hero, having played hockey for a Junior B affiliate of the Montreal Canadiens and starred in high school basketball (he later barnstormed with the Harlem Globetrotters). In baseball the limited Ontario schedule restricted his starts to seven or eight in each of three seasons of amateur ball. Those looking for clues as to his extraordinary ability to pitch so many major league innings may refer to such a sheltered upbringing as cause.

Fergie debuted for the Phillies in 1965 and his first pitch put Dick Groat on his back. Pat Corrales, his catcher that day, recalled, 'The next three deliveries were on the black.' Jenkins later criticized such tactics. 'I figure if a pitcher wants to go headhunting,' he told a *Sports Illustrated* reporter, 'he should play hockey instead of baseball.'

Jenkins made his reputation on the strength of pin-point accuracy and a slider he figured he could get in the strike zone 80 per cent of the time. At his peak he averaged around 1.8 walks per game. Statistics compiled by United Press in 1987 showed that teams giving up more than three walks had a winning percentage of .487, but for two walks or less the percentage climbs from .582 to .659. And while his Canadian fans were disappointed at the abrupt conclusion to his career before the 1984 season just 16 victories short of 300, it did guarantee him status as the only pitcher to throw over 3,000 strikeouts (3192) against fewer than 1,000 bases on balls (997). The downside to such precision was that as a flyball pitcher he gave up a lot of sacrifice flies and a career total of 484 home runs. Only Robin Roberts at 505 leads him in this dubious category.

His teammates seldom complained. In 1974 Texas Rangers' rightfielder Jeff Burroughs said, 'Before Fergie it used to get boring in the field. Our pitchers

would walk so many hitters and get behind so many others, you'd lose your concentration. Fergie is always around the plate, so you have to be alert.'

Nowhere was this better appreciated than during six glorious seasons with the Cubs between 1967 and 1972 when he always won twenty or more games and the Cubs had three seconds and three thirds. Before a game on July 12, 1969 against the Phillies, his manager Leo Durocher told Tony Kubek that 'Fergie's our stopper like Ford was on your club. He believes he's going to beat you.' He will be remembered for his familiar loping run to the mound, his economical delivery, and how he placed his legs in a good square position to field after the pitch. His 363 career putouts are a modern major league record for pitchers. He contributed at bat as well averaging .165 and hitting 13 career home runs. In the 1969 season Jenkins started 42 games, more than at any other time in his career. He completed 23 of those starts and led the league with 273 strikeouts. Yet he surpassed that season's 311 innings pitched in each of the next two seasons (313 in 1970, 325 in 1971—his Cy Young year).

Jenkins was a player best respected for his career rather than his peak value—a pitcher who performed best over an entire season but in the truly big game (as rarely as that occurred in a career which did not include a pennant winner) was subject to greater calamity than a true power pitcher might be. When the Cubs collapsed in September 1969 before the onrushing New York Mets, Fergie was as guilty as anyone, as he recounted in his autobiography.

> I wanted to be Doug Harvey [Montreal Canadiens All-Star defenceman in the 1950s and, ironically a minor league baseball player from 1947-50 with Ottawa of the Border League where he averaged .344]. Doug Harvey was my idol.
>
> Ferguson Jenkins

> I was to blame as much as anyone else. I lost three starts in a row in early September before winning my twentieth game, and even that was not a good performance. I struggled into the eighth inning against Philadelphia on September 17 before I was relieved. I had given up seven hits and four walks, and my hitting was better than my pitching. I cleared the bases with a three run triple in the eighth to break a 3–3 tie, and we went on to win 9–7.

After his major league career ended Jenkins played a season in Ontario's Inter-County League in 1984 with the London Majors and in 1989 pitched in the Florida based Senior Professional Baseball Association. In the space of a week in January 1991 Jenkins won a place in baseball's Hall of Fame and lost his second wife to injuries sustained in a car crash. His life assumed Job-like proportions with the death of his daughter, but he remains a powerful image of Canadian sporting achievement.

Other notables in Ontario baseball history

Babe Ruth, the greatest player in baseball history, visited Canada often. As a rookie pitcher with Providence he hit his only minor league home run at Toronto's Hanlan's Point ballpark on 5 September 1914. He would return to Toronto at the peak of his career for an exhibition game as a New York Yankee in the twenties. He also went to Quebec which he had first visited as a Providence ballplayer, and at an exhibition game before 4,000 fans at the Guybourg grounds he hit a home run which some claimed travelled 600 feet.

Ruth also visited Vancouver and Nova Scotia for baseball-related events during his life but the most significant Canadian connection was the man he credited as the most important influence on his life, Brother Matthias of St Mary's Industrial School in Baltimore. Babe spent most of his boyhood there and Matthias who belted out balls with a bat in one hand introduced the future slugger to baseball and attempted to marshal Ruth's potentially destructive behaviour with positive direction. Matthias's name before entering the Order was Martin Boutlier. He was born in Lingan, Nova Scotia (in Cape Breton) in 1872 and played the game while growing up there before eventually joining the Xavieran Order and moving permanently to the United States. Matthias taught Ruth to both hit and pitch, a product perhaps of Boutlier's own baseball upbringing in Canada.

After his playing career Ruth lived with the hope of a big league managing job that never came. During one of those years of hope he vacationed in Nova Scotia. Arriving in Yarmouth aboard the SS *Acadia* on 4 July 1936 he motored through the Annapolis Valley, played golf, and on 7 July put on a batting exhibition during a break in a game in the mining town of Westville between the home-town Miners and the Liverpool Larrupers. The Canadian media's ongoing fascination with his major league career helped elevate baseball and Ruth to mythic proportions for an entire generation of fans.

Perhaps the greatest hockey player of all time, Wayne Gretzky, also has an important Ontario baseball connection. On 3 September 1973 the Chatham Kinsmen Peewees won the Canadian National Beaver baseball championship in North Battleford, Saskatchewan. To get there they had to beat a select team from Quebec who were sponsored and outfitted by the four-year-old Montreal Expos. The Chatham team was allowed to pick up players from other teams and one of those was Gretzky, a small, skinny 12-year-old from Brantford. He bunted his way to a .500 batting average against Quebec and Chatham beat them two games to one. Out west they outscored their opposition 81–11 and in the championship game blanked Sydney, Nova Scotia 8–0. Gretzky eventually became one of Canada's greatest hockey players. Because injury kept him from playing in the autumn of 1992, he was in Atlanta's Fulton County Stadium to see the Blue Jays win the World Series.

Final observations on baseball and politics

After his major league baseball career Fergie Jenkins returned to Canada and had an unsuccessful run for political office in the Province of Ontario as a member of the Liberal Party. The affiliation is loaded with significance. There is a curious parallel between baseball enthusiasm and Canada's Liberal Party throughout the country's history.

In the 1870s William Climie of Bowmanville struggled to introduce baseball against the cricket partisanship of his Conservative Party rival Frederick Cubitt. Erastus Wiman supported Liberals and baseball, and this connection continued in the person of Pierre Trudeau's father. Perhaps the most notable baseball role was played by Liberal Party leader and Prime Minister Lester Pearson. He had played baseball as a youth in Guelph and later as a diplomat in Washington. During his time as the country's leader in the 1960s the game was in decline but Pearson would regularly preempt cabinet meetings in the early fall to watch that year's World Series games on television. Baseball's revival corresponded to the Montreal Expos entry into the National League in 1969 and appropriately Pearson acted as the club's honorary president in the last years of his life.

Opening day in Toronto in the late 1950s; an Ontario Lieutenant Governor and a Cuban revolutionary make for an odd couple

Victorious Toronto Maple Leafs celebrate 1954 International League pennant. From left, back row: trainer Bill Smith, Billy DeMars, Frank Barnes, Sam Jethroe, Ray Shore, Archie Wilson, Elston Howard; middle row: Eddie Blake, Charley Kress, Ed Stevens, Loren Babe (waving), Bob Weisler, Connie Johnson, Fred Hahn, Harry Schaeffer, Buddy Kerr, Rudy Minarcin, Arnie Landeck, Don Griffin; front row: Manager Luke Sewell, Lew Morton, Vic Lombardi, Hector Rodriguez, coach Bruno Betzel, Mike Goliat, and Jack Crimian applying the shampoo. York University collection and Russell Field

It is ironic therefore that one of the driving forces behind the awarding of an American League expansion franchise to Toronto was a well-known Conservative politician, Paul Godfrey, who as Metropolitan Toronto's chairman in the 1970s convinced another Conservative, Ontario Premier Bill Davis, to upgrade Exhibition Stadium to major league baseball standards. This was a politically risky decision but it was rewarded in 1976 when the American League voted to expand into Toronto for the next season. For several years Toronto and Montreal played an annual Pearson Cup game to determine Canadian major league bragging rights, but that custom has since been abandoned. Pearson's and Godfrey's baseball enthusiasm are proof that politics and baseball can occasionally be nonpartisan.

The evidence of such nonpartisanship would be demonstrated in 1992 and 1993 as a Canadian baseball team pursued World Series glory and the whole country joined the celebration.

CHAPTER ELEVEN

Toronto Blue Jays

Part One
World Series Champions

The two World Series victories of the Toronto Blue Jays have sparked thoughtful reflection on the nature of their Canadian identity. One Canadian television commentator even quipped, following the 1992 World Series, that our Americans had beaten their Americans.

Drawing conclusions based solely on the birthplaces of ballplayers misses the entire point of professional sport. A team's makeup is a matter of commercialism, not nationality. Today's teams differ little from comparable 19th century squads of itinerant professionals throughout North America who first forced local amateurs off their home town squads. For sports fans of the last century and this, the crisis of this relocation was resolved by a willing suspension of disbelief that these players could somehow represent them. The success of this suspension of disbelief was expressed by a character in the play *The Tramway End*, by Dermot Bolger. Of his country's soccer team made up of exiles, and sometimes second- and third-generation Irishmen, he says they were 'the only country I still owned, those eleven figures in green shirts, that menagerie of accents pleading with God.'

In the end people who attend games care less about the civic or national origins of players than their level of performance. Players have been deliber-

> One reason outfielder Mitch Webster was happy to be a Chicago Cub:
> 'My entire major league career was in Canada. It's great not to have to listen to two national anthems.'
>
> *USA Today*
> 5 August 1988

ately cast as mercenaries to allow spectators to discard those who can no longer perform—there being less emotional commitment to those who are neither one's neighbours nor fellow citizens. In the world of professional team sport the entire question of the national origin of Blue Jay players is irrelevant.

In three other ways however the Blue Jays could be said to have a distinct Canadian identity. One is in the management/ownership of the Blue Jays in the person of President Paul Beeston and the corporate ownership of Labatt Breweries. For his part Beeston acknowledges the influence of the Montreal Canadiens hockey team as a model for the modern professional sport franchise. It is a model which respects the past in its regard for the team's history and players, and the future in its understanding of the elements of successful team management. In the case of baseball, Beeston argues that management is required to use player development, trades, and free agency to build a team, and that the only reason to operate a sports franchise is to win. In this the Blue Jay philosophy conforms to the team-building tactics of the Canadiens' one-time manager Sam Pollock who now sits on the Blue Jays' board. There was added acknowledgement of the success of the Blue Jay way in *Sporting News*' designation of Pat Gillick and Cito Gaston as co-winners of its 1993 Man of the Year award.

There was an equally significant Canadian aspect to the way Toronto sports fans responded to the World Series victories, not by trashing their downtown as had become a North American custom, but in a public ceremony marked by politely enthusiastic behaviour. It was behaviour in other words that was a genuine reflection of the game's placid character as represented in its most endearing literature from Kahn to Kinsella, Angell to *Casey at the Bat*. Toronto has been criticized for using seemingly endless resources to buy a winning team. Winning by legitimate means should be the purpose of all professional teams and by supporting their team in record numbers Toronto's fans effectively conspired with the team's owners to provide the resources to make winning possible. All successful franchises depend on such willing civic conspiracies.

Who would argue against the notion that there are few greater joys than the pleasure of victory removed from any financial reward? This is the third element of the Blue Jays' Canadian identity. It was one willingly entered into by millions of Canadians who, like Bolger's soccer fan, chose to identify with these pursuits of glory and ultimate victory. This is no small accomplishment in a country which often reviles anything with a Toronto logo.

Canadians expressed some disappointment at American apathy to the win, but to the extent that some Americans ignored this story it could be said they were denying themselves a part of their own baseball history and therefore creating a wilful blind spot. Such censorship ultimately harms the censor for it forces him to acknowledge that there are items that even with his power he must fear and keep from public disclosure.

The road to victory

On Wednesday, 22 April 1977, the Toronto Blue Jays' cumulative regular season win-loss record dropped below the .500 mark, not to resurface again at .500 until 1 September 1993. In between they had won one World Series, one American League pennant,

and four divisional pennants. By the end of 1993 their season percentage had improved to .501 based on 1351 wins and 1344 losses and they had added another World Series, and of course additional American League and Divisional pennants. With the creation of an extra division and an additional round of playoffs for the 1994 season, the era commencing with the Blue Jays entry into the American League as an expansion team in 1977, had come to an end.

Joe Carter's dramatic ninth inning home run aside, the World Series victory of 1993 was confirmation of what 1992 had wrought. It was indeed possible for a Canadian-based team to win the World Series. It provided a spark for national celebration and an opportunity for reflection on both the history of the game in the country and its powerful grasp on the everyday lives of many Canadians.

> I felt proud to be a Canadian. I felt myself trying harder to do something in Toronto.
>
> Dave McKay, first Canadian with Toronto Blue Jays (later strength coach with Oakland)

Brief history

On 26 March 1976 the American League had awarded a franchise to a group consisting of Imperial Trust, Labatt's Breweries, and the Canadian Imperial Bank of Commerce after the consortium had been rebuffed by the National League in their attempt to purchase the San Francisco Giants and move the franchise to Toronto. At the time the American League was under legal challenge to place a team in Seattle whose 1969 expansion team had been moved to Milwaukee after only one season in the Pacific Coast city. Toronto's timely pursuit of a major league franchise allowed the American League to attain a perfect east-west balance by admitting both Toronto and Seattle.

Paul Beeston was the new organization's first employee and among the crucial early recruits was Pat Gillick as Vice President of Player Personnel. The Jays' home was Exhibition Stadium, a converted football facility, where they began play in a snow storm in 1977. For the next six seasons the Jays registered losing seasons. In 1983, however, they started a run of consecutive winning seasons lasting through their World Series winning years. The Jays won their first divisional title in 1985 but missed a trip to the World Series after blowing a three-games-to-one lead to Kansas City in the American League playoffs.

The Blue Jays' first Canadian ballplayer, Dave McKay in 1977. Toronto Blue Jays

Damaso Garcia, early Blue Jay star at second base. Courtesy David Crichton

Another East Division title followed in 1989, the year they moved into SkyDome, but this time they lost to Oakland in the playoffs. One winning series away from the World Series in 1991 they again lost, this time to Minnesota. Determined not to experience such a fate again the Jays signed free agents Jack Morris and Dave Winfield in the off season and prepared for the 1992 season.

1992

Within the span of baseball history that stretches from the first games in northern Europe ten thousand years ago to Joe Carter's season ending home run, the ultimate moment for the fan is the moment his team first wins a World Series—when cheering for the home team is at last vindicated. One will always be able to pick up a *Baseball Encyclopedia* and see it recorded.

In a late March 1992 spring training game in Florida with two out in the ninth, a two-strike count on Jays batter Jeff Kent, and the Jays trailing 3–2, the promising utility infielder hit a three-run homer to give Toronto a victory. At the same time in New York City, the *Daily News* reported that Mets' pitcher David Cone was about to break his 'cone of silence'. A bullpen incident involving Cone and some female minors would hasten Cone's eventual departure from New York.

The Blue Jays' first manager Roy Hartsfield demonstrating his sliding technique at spring training in 1978. Toronto Blue Jays

Toronto opened in Detroit and their expensive off-season free agent acquisition Jack Morris took the mound for his 13th first-day start. Despite some late inning troubles, he pitched the Jays to victory. Early in the game Pat Borders was accosted on the field by a notorious fan, the busty Lulu Devine. He attempted to ward off the threat of her kiss with a bat and afterwards responded by hitting a home run. Meanwhile Dave Parker, who played for the Jays during the last month of the 1991 season, had not found a willing buyer for his services. He said baseball was like a third world country in which there was room for only the very rich and very poor—implying that middle range players in talent and salary were being eliminated.

At the Jays 1992 home opener Toronto was down 3–2 in the ninth when Pat Borders, still recovering from the Lulu episode, homered to tie the game. With two out Devon White doubled, and Robbie Alomar singled him home for the win. Alomar, whose home was the hotel in the SkyDome's centre field, said that he separated himself from the game by having a suite that didn't look out onto the field.

Despite a fast start of five wins, White said 'We still have 157 to go.' The Jays' disappointing loss in the 1991 American League playoff had caused many to question the team's fortitude. The word choke had even been used. Advice was available from odd sources. A circus performer described how he overcame the gagging response by systematically sticking a coat hanger down his throat three times a day for five to six years.

The April 20th match at Fenway Park began at 11 a.m. in deference to the running of the Boston Marathon. It proved to be a marathon itself, going 13 innings before the Jays won. Robbie Alomar was having a torrid start. While one can speculate on the different styles and body types of those who hit singles, doubles, triples and homers, Alomar seemed to encompass all features in one person. In early May he had a run of seven straight hits and nine consecutive at-bats in which he reached base. Each achievement was one behind the team record of Rance Mulliniks, the durable third baseman in his last season of play.

At the start of the 1992 season six players with 16 years or more service in the majors had yet to win a World Series. They included Robin Yount, Dale Murphy, Carlton Fisk (all of whom had retired by the 1994 season without a ring), Andre Dawson, Charlie Hough, and Dave Winfield. Winfield had been the

Fred McGriff, part of trade that brought Joe Carter and Roberto Alomar to Toronto. Courtesy David Crichton

Jays' other key off-season free agent signing and had been cruelly dubbed 'Mr May' by New York Yankee owner George Steinbrenner for his failure to produce in the 1981 World Series. Another leading player of the eighties without a ring was Toronto's one-time ace pitcher Dave Stieb and, after coming close on three occasions, the only Jay to throw a no-hitter. Now injury-prone he nevertheless turned in a stellar performance against Milwaukee in early May.

The International League which had not had a Canadian presence since the Winnipeg Whips played in the early 1970s announced the awarding of an expansion franchise to Ottawa. Meanwhile the Mets' David Cone came close in his attempt to be the first Met to throw a no-hitter (the Padres were the only other team without a no-hitter in their history). Canadian ball fans now had American superstations available on cable television and could witness play from Wrigley Field before the ivy comes out in the spring.

Much of the early sock in the Jays' line-up was coming from Alomar, Carter and Winfield. On 7 May the Jays fell behind the Seattle Mariners 6–0 and entered the ninth down 7–3. They scored one and loaded the bases for Winfield who hit a grand slam off Mike Schooler for the eventual win. In late May, George Bell, the only Jay to win an American League Most Valuable Player Award (in 1987), returned to town in the uniform of the Chicago White Sox. With the bases loaded in the first Juan Guzman stuck out Bell and the Jays went on to win 3–0. The bullpen had had a tendency to blow leads they inherited from Guzman; a popular theory was that he was so strong the bullpen looked less intimidating—an interesting speculation given the presence of set-up man Duane Ward and the terminator Tom Henke.

A few days later the Twins razzed Alomar for not running out ground balls and Todd Stottlemyre for losing his cool after giving up a few hits. Earlier in the season he had almost blown an 8–1 lead to the Red Sox and had to be removed after four innings.

White answered their catcalls on 1 June by leading off a game with a home run batting righthanded, and then broke a tie in the tenth inning with an inside-the-park home run batting lefthanded. He was only the sixth player in major league history to combine a leadoff and extra inning four-bagger. Billy Hamilton had been the first to do so in 1893. White was also the 56th player to hit home runs from both sides of the plate in a game.

Jays manager Cito Gaston whose job, threatened by the Jays' quick exit from the 1991 playoffs, was saved apparently by the intervention of club president Paul Beeston, said that Alomar might be the first Hall of Famer to wear a Blue Jay uniform. Meanwhile Jay utility infielder Alfredo Griffin's hit on 27 May left teammate Rob Ducey as the only hitless player in the league.

A whimsical piece on CBC radio suggested that a city might be defined as a place where you can't be accosted by wild sheep. On such a note the Jays went into Yankee Stadium in early June and Candy Maldonado hit a tremendous home run to the deep-

Tony Fernandez, member of 1993 World Series champions

Jimmy Key, member of 1992 World Series champions

Dave Stieb, member of 1992 World Series champions

Jesse Barfield

est centre field bleachers. It was several minutes before New York's rapacious fans could reach the ball.

By mid June, boos greet Kelly Gruber who remained in the fifth spot in the line-up despite woeful production and his complaints about undiagnosed injuries. An ominous closed-door meeting was held between Gruber and the Jays braintrust of Paul Beeston, Pat Gillick and Cito Gaston. Ironically George Bell was reported to have said that Toronto fans only booed Latins and blacks. More likely targets however seemed to be Jays who wrote books, Gruber having joined a literary roster that included among others Dave Stieb and Bell who, when they played poorly, both heard negative choruses from Toronto's normally docile fans.

By late June only Guzman and Morris seemed capable of winning. A local broadcaster reminded Jays fans that as Casey Stengel once said 'You can't win all the time so lose right.' Bearing this out was Toronto southpaw Jimmy Key who performed valiantly despite an absence of positive results. Opposing pitchers had an earned run average of 2.76

when facing Key and 3.77 when facing any other American League pitcher. The Jays were reported to be pursuing a quality pitcher but a trade for the California Angels' gifted lefthander Jim Abbott fell through. Toronto had drafted Abbott several years before despite the fact that he was born with an only partial right arm but the pitcher decided to go to university instead.

On Canada's July 1st birthday Greg Myers doubled home Derek Bell for a Jay win. Myers left the Blue Jays shortly thereafter along with the team's only home-town boy Rob Ducey in a trade with the Angels for Mark Eichhorn. Before they went however the Jays made nine straight hits against the 'Halos' on 3 July and two days later delighted their sundrenched fans on bandanna day with a solid 6–2 victory, though pitcher David Wells complained that fans booed him after he surrendered a meaningless home run. Wells was neither black, Latin, nor an author.

In mid July, Blue Jay fans listening to broadcasters Tom Cheek and Jerry Howarth might have assumed the game was being played at SkyDome.

George Bell, American League's Most Valuable Player in 1987

Tom Henke, member of 1992 World Series champions

Thousands of fans from British Columbia came down to Seattle to cheer on the Jays and were rewarded as the team reacted to Cito Gaston's ejection for arguing a ground rule double call by sending eleven players to the plate. Bob Bailor, often mentioned as Gaston's eventual successor, got the managerial win. Across the continent in Syracuse, New York the Blue Jays' Triple A farm team, the Chiefs, were rained out on car give-away night and the players lobbed water bombs into the box seats with a homemade catapult. A somewhat testy Ed Sprague reluctantly signed a few autographs, perhaps wishing he was in Seattle.

On 20 July the Red Sox introduced Paul Quantrill, the 187th Canadian to reach the majors. Blue Jay fans from Bowmanville to Port Hope recognized the name, as his father owned one of the largest car dealerships in the region. Elsewhere on July 23 the Jays ace Juan Guzman left the game with soreness in his right back below the shoulder. Poetically the winning pitcher that night was Dave Stieb. It was his last ever victory as a Blue Jay.

Guzman had been traded to the Jays for Mike Sharperson in 1987 when the Dodgers despaired of his ability to find the strike zone. Slowly Guzman learned the lesson of veteran pitchers like Rick Sutcliffe who said he threw strikes to get ahead in the count and then balls to get the batter out.

Guzman briefly appeared on 3 August in Boston. The Red Sox' Billy Hatcher stole home and Gaston and Guzman had an ugly disagreement on the mound. Guzman later said he couldn't make certain throws and was placed on the fifteen-day disabled list. In his absence the Jays rode Jack Morris's arm through a rocky August. Between 1 August and a 3 September off-day, the Jays' lead tumbled from four and half games to just a half, but Morris won six important games.

Doug Linton lost his first game as a Jay on 8 August. When the Jays lost the next day it marked a period of eight games in which the starters had a combined ERA of 10.38. The season's turning point may have been a 13 August afternoon game with the

Baltimore Orioles. On the verge of a three-game sweep which would have moved them into a first place tie with the Jays, the Orioles figured they had easy pickings with the rookie Linton on the mound. He was the ultimate long shot. Only one out of 20 players drafted ever makes the bigs and as a 43rd round pick in 1986 Linton's odds were much worse. After a spectacular 14–2 debut with Myrtle Beach in 1987, he had been a .500 minor league pitcher. Against Baltimore however his gutsy eight inning performance gave the Jays a 4–2 win and gave him a solid right to a future World Series ring though he didn't pitch with Toronto after 29 August.

At the Olympic Games a powerful American squad did not win the gold medal and Jeffrey Hammonds, who had been drafted by the Blue Jays in 1989 but did not sign and was later selected by the Baltimore Orioles, said if he had known the result he doubted he would have spent his time with the Olympic team. Nationalism was in greater evidence among over 10,000 Jays fans from Manitoba and Saskatchewan who invaded Minnesota for a late August series. Pitcher Jack Morris avoided the Jays' hotel which had been besieged by Canadian fans.

The answer to the Jays' pitching woes was a spectacular trade on the last weekend of August as Toronto sent minor leaguer Ryan Thompson and the very able utility infielder Jeff Kent to the New York Mets for righthanded pitcher David Cone. 'You don't get too many chances to win it all,' Detroit manager Sparky Anderson said. 'So who cares if Cone walks away after the season is over.' At the time the Jays were 34–13 when Guzman and Morris started and 37–40 behind their other starters. Before Cone's Saturday, 29 August start however the Jays were humbled the night before by their closest September pursuers, the Milwaukee Brewers. Not only did the Brew Crew win 22–2 but almost became the first American League team to score a run in every inning, being blanked only once.

On the 4 September night that Cone won his first game as a Jay, Dave Winfield made his famous 'Winfield wants noise' speech and Toronto fans responded as their team scored 16 times. On 5 September Guzman won his first game since just after the all-star break and the next day the former Syracuse Chief Ed Sprague hit a three run homer as the Jays nipped the Twins 4–2. Scouts had called Sprague the best situational hitter in the organization, a judgement full of meaning in the second game of the 1992 World Series.

By 15 September the Jays had gone up by four but fans' palms had gone permanently clammy with apprehension. Pennant races are meant to be enjoyable, but fans of the contending teams experience feelings ranging from exhilaration to white-knuckled fear. The Brewers stoically pursued the Jays and their respected hitter Paul Molitor who was to become a free agent at season end said he would be interested in playing only in Milwaukee or Minnesota. It was noted at the time that Winfield might be the most dangerous hitter in baseball because of his ability to adjust to circumstances. In September he became the oldest player to get 100 RBIs. Another veteran Jay, Alfredo Griffin, who first played for Toronto in 1979 before leaving in a trade after the 1984 season, had returned this season and said he cheered for the Latin kids. Alomar noted his respect for the aging infielder.

In September a Canadian big leaguer of the 1940s, Aldon Wilkie from Saskatchewan, died. Wilkie had been part of an ill-fated strike attempt by members of the Pittsburgh Pirates seeking better playing conditions, conditions that today's players take for granted. The action probably hastened Wilkie's departure from the majors in 1946.

On 17 September John Olerud's two-run homer in the tenth continued a Jays streak which lasted to season end. After 29 August the team never lost two in a row. 'At a certain point in your career, winning outweighs everything else,' Joe Carter said. To relieve the pennant race pressure backup catcher Mike Maksudian revealed a penchant for eating live locusts. Meanwhile Larry Hisle was looking for a way to speed up the elaborate computer program he wrote himself to track 20 variables (including speed, weather, and location) for each pitch the Jays' hitters faced. For some commands Hisle's retrieval time was five minutes. A software producer offered help. 'I'm confident the system will be operational by the play-offs,' Hisle said.

In New York on 27 September Jack Morris became the first Jay pitcher to ever win 20 games in a season but he had to wait out a fifth inning rain delay with his team up by nine.

On the last weekend of the regular season, Jays fans were mulling over the question whether Duane Ward or Tom Henke should complete games. On Friday night the Jays did not clinch the pennant because, in a harbinger of things to come, Oakland's ace reliever Dennis Eckersley failed to hold an Oakland lead against the Brewers. The next day in a reversal of usual roles Ward relieved Henke as the Jays won the American East pennant and a healthy Guzman earned his 16th victory. In the process the Jays had become one of only six teams in the 20th century that had not been swept in any season series (the others being the 1943 Cardinals, the 1921 Indians, the 1910 Cubs, the 1905 Athletics, and the 1904 Giants).

At this point the Jays' ten-year record from 1983 through 1992 was 906–711, putting them 36 games ahead of Oakland in the decades standings. Toronto however had yet to win an American League pennant and a trip to the World Series. Against the Athletics it appeared that their American League playoff series would be tied at two games apiece when Dennis Eckersley surrendered a game-tying two-run homer to Robbie Alomar. The Jays won in extra innings but were stymied in game five by, ironically, their future 1993 American League playoff hero Dave Stewart. The achievement which had eluded the Jays in 1985, 1989, and 1991 was theirs in game six as the Jays scored early and often, leaving their fans to wring their hands until Maldonado caught the final out.

After losing game one of the World Series to the Atlanta Braves, Ed Sprague's pinch-hit home run turned the series around in the ninth inning of game two. The events of that game were fraught with meaning for Canadian sports fans. Sprague's wife had won a gold medal in synchronized swimming earlier that summer when a Brazilian judge incorrectly reported a mark for Canadian Sylvie Frechette, knocking her into second place. (Frechette was later awarded a co-gold medal with Kristen Babb-Sprague.) At the time Sprague implied he didn't really care what Canadian fans thought. Now he had become their hero. More significantly, before game two US Marines had flown the Canadian flag upside down, prompting a national furore and demonstrating that this event was far more than a contest between two teams of highly paid imports. Game three marked the first time the World Series had ever been played in a locale outside the United States. Toronto fans cheered both the Canadian and American anthems and then witnessed Devon White's spectacular catch against the centre field wall to stop an Atlanta attack and start a double play in which the umpire missed a third out.

The Jays went up three games to one the next evening but a chance to win the World Series at home was derailed by Lonnie Smith's grand slam home run off Jack Morris in game five, causing Morris to quip that the Jays were sure to win now because he would not pitch again in the Series. On a Saturday, 24 October game that went into Sunday, the Jays were one strike away from victory in the ninth when Tom Henke gave up a game-tying RBI hit by Otis Nixon. In the eleventh inning Dave Winfield drove in two runs. The Jays needed both, as once again Nixon came to bat in the bottom of the eleventh with the Braves down by a run, two outs, and the tying run on third. Reliever Mike Timlin scooped up Nixon's drag bunt and threw him out at first.

Canada exploded in the moment that Joe Carter leaped off first base.

There comes a point in the fans' private agonies and tribulations when they imagine that they become a factor in a team's success. If that is the case Blue Jay supporters who surely stretched from one end of the country to the other—a nationalism of which Dave Winfield acknowledged he was aware—had finally willed the unthinkable, a World Series victory by a non-American team. Jays had only 35,000 unsold tickets for the entire season. More important, they had finally reached a franchise pinnacle.

A year later it happened again at home and among those celebrating was World Series Most Valuable Player Paul Molitor who had opted to join the Jays because of their explicit dedication to winning. For Canadians who had sometimes shown a willingness to applaud good losing efforts, the pleasure of a double victory had an almost decadent twist of guilt and thrill mixed into one celebration.

Part Two

The Jays do it again

'Don't ever change,' David Cone urged Toronto fans and, by extension, those throughout Canada at the grand SkyDome celebration two days after the team's World Series victory. Cone had caught the spirit of the victory as one in which all Canadians shared. It was somehow a national feast capping a century and a half of baseball enthusiasm.

But the winds of change were quickly dismantling the 1992 Blue Jays. On the very day of the victory party, Mike Maksudian was claimed on waivers and the Jays front office was privately informing Dave Stieb that he did not fit into their 1993 plans. Free agency had set many long-time Jays adrift, among them Tom Henke, Manny Lee, and Jimmy Key, while it was doubtful that either Dave Winfield or Joe Carter, another free agent, would return.

Back in 1985 George Bell had argued that apparent umpiring mistakes in that year's pennant series with Kansas City were the result of anti-Canadian feelings. 'If our ball club was American,' he was quoted, 'everything would be fine . . . the series would be over.' Such paranoid sentiments had a certain currency among many Canadian ball fans until the win of 1992 and even for a few days afterwards, perhaps harking back to Ben Johnson's loss of his 100-metre gold medal several days after crossing the finish line first at the 1988 Seoul Summer Olympics. Nevertheless the feeling persisted that the 1992 triumph was possibly a once in a lifetime event and line-up changes would scuttle any Blue Jay and therefore Canadian hopes of repeating it.

Soon after season end the Blue Jays lost another prospect: in the expansion draft Canadian-born Nigel Wilson was the first draft choice of the Florida Marlins (though he returned to Canada in 1993 as a member of the Marlins' Triple A affiliate in Edmonton). As they entered the winter meetings in December the Blue Jays' number one goal was to re-sign Joe Carter. In the coming seasons events surrounding that fateful week of free agent signings would profoundly influence Blue Jay team fortunes for much good and at least some ill.

Carter has always been a proven run producer with the ability to hit 30 home runs and drive in 100 runs. Losing him would have changed the team's entire policy. It would have meant finding a power hitting replacement, and the pursuit of Paul Molitor would likely have been abandoned. Molitor's ability to get on base would be blunted by the absence of someone who could get him across the plate. Failure to replace Carter would most likely have caused the Jays to abandon hopes of repeating and instead started a re-building process, one that could take several years to pay dividends.

Compounding the Jays' pursuit was the interest of the Kansas City Royals in Carter who was building a new house in that city. Carter weighed both offers

The Blue Jays All-Star second baseman Roberto Alomar, hero of the 1992 American League playoffs against Oakland. Courtesy Howard Starkman, Toronto Blue Jays Baseball Club

and relied eventually on a form of divine intervention. He would later claim that a dream in which he saw himself playing right field for Toronto alongside Devon White was the turning point in his decision. Getting Carter back freed the Jays to complete a deal with Molitor who had played his entire 15-year major league career with the Milwaukee Brewers. When Jimmy Key opted for a four-year contract with the New York Yankees, the Jays signed another veteran, Dave Stewart.

The Blue Jays of 1993 offered an odd assortment of even greater achievement than the year before mixed with occasional and inexplicable mediocrity and near tragedy. On the positive side John Olerud emerged as a major league star and, until slumping somewhat in August, had a chance to be the first major leaguer, since Ted Williams in 1941, to bat .400 for the season. Blue Jay fans, at first bitter at the team's refusal to meet Dave Winfield's contract demands and his resulting departure, were then overwhelmed by the marvel of Paul Molitor's clutch hitting ability. At second base Roberto Alomar continued to make the outstanding look normal. At season end Alomar trailed only Olerud and Molitor in the batting race—the first time three players from the same team had led this category since Philadelphia's 1893 National League team.

Other seasonal highlights included the All-Star game in Baltimore where Cito Gaston's line-up included not only Olerud, Molitor, and Alomar, but Carter, White, Duane Ward, and pitcher Pat Hentgen who would later come within a victory of becoming the second Jay to post 20 wins in a season. Gaston was subject to vociferous heckling from local Oriole fans who thought he should have brought in Baltimore pitcher Mike Mussina rather than his own reliever Ward to pitch the final inning.

During the three day All-Star break the Jays' third base coach Rich Hacker was seriously injured in an automobile accident. A third base coach is a wonderful testament to the powers of sign language; it's his job to relay signals across the diamond to runners or the hitter. A runner on first, for instance, needs to know whether the batter is going to let a pitch go by or try to make contact. Hacker's recovery was slow and though he eventually returned to the team he was replaced at third by a new signaller, Nick Leyva.

Baseball is a game of unspoken communication and nowhere is this more so than between catcher and pitcher. A 19th century Guelph newspaper once

referred to 'a system of freemasonry between the catcher and pitcher, as they were continuously manipulating their fingers at each other'. Throughout the season catchers Pat Borders and Randy Knorr struggled to communicate with Juan Guzman over pitch selection, the pace of his delivery, and his general tendency to throw balls in the dirt. By season end he had delivered a 14–3 record; his total over three seasons of 40–11 was one of the best career starts for a pitcher in baseball history.

The Jays had started their season slowly, losing in Seattle 8–1. By 12 May not only was the team four and a half games out of first with a 16–17 record but their starting shortstop, Dick Schofield, was lost for the season after falling awkwardly while defending second base. On 10 June the Jays took a page from their own history, re-acquiring Tony Fernandez in a deal that sent disappointing Darrin Jackson, who in spring training had himself been traded for (Derek Bell going to San Diego). Jackson's only Jays' highlight had been an early season solo homer in a 1–0 win over Chicago. Fernandez paid immediate and ongoing dividends. Against Detroit for his first weekend series, he drove in five runs in a Sunday game, sparking a 13–3 streak by the team and their return to first place. Fernandez and the Jays had been cruelly robbed of a post-season berth in 1987 largely because Tony had been injured after a collision and fall at second base in the last two weeks of that season. Now he and Alomar became pictures of grace in the centre of the infield.

Joining the Jays about the same time as Fernandez was Toronto-born Rob Butler. His spray hitting style and fleetness in the outfield impressed the home-town crowd. Butler's season was interrupted on 22 June when he injured a thumb sliding into second.

Both of Canada's major league teams pay special attention to local recruits. Larry Bearnarth, a former coach with the Montreal Expos, once remarked that 'I'd sit up and take notice in meetings when it was mentioned this kid or that was from Sarnia or Stratford. No one would ever say this guy's from Texas, or Ohio, and New York. Besides, Canadians had more guts.' While the Jays with record attendance had no need to play Canadians for their home-town appeal, the team has always maintained its own Canadian scouting office, one originally directed by Bobby Prentice.

The Jays' winning ways quickly subsided and

there began an incredible period lasting up to the All-Star break during which they lost 10 of 11 games. It was a spell with ominous lapses in pitching and particularly hitting. Through July the Jays remained in a tight pennant race. Feeling the need to shore up their offence and outfield the Jays dealt for talented Oakland star Rickey Henderson but had to give up a highly rated pitching prospect, Steve Karsay. Entering an off-day on 2 September they led the American League East by two and a half games and then, at a time when good teams make their run to the finishing line,

Paul Molitor, 1993 World Series Most Valuable Player.
Courtesy Howard Starkman, Toronto Blue Jays Baseball Club .

the Jays proceeded to lose six in a row and fell back into a tie for first.

The Jays seemed to say—like the weather in Western Canada—if you don't like our performance now come back in an hour. In the following 17-game span the Jays won 15. During this run *USA Today Baseball Weekly* ran a cover story on the 'Damn Blue Jays'. A sub-heading in the story, in mock seriousness, described the Jays as 'Rich Foreigners'. In the American baseball public a nerve had been struck.

The Jays clinched the pennant in Milwaukee and a few days later headed to Chicago for the American League pennant series. Michael Jordan threw out the first ball and shocked local fans later that evening by announcing his retirement from basketball. He would resurface in the minor league baseball uniform of the Double A Birmingham Barons of the Southern League in 1994.

Led by Dave Stewart's stellar pitching the Jays outlasted Chicago in six games. The Jays returned home to face the Philadelphia Phillies in the World Series. They traded wins in Toronto. In game two the Phillies' mercurial reliever Mitch 'Wild Thing' Williams guessed correctly that Alomar would try to steal third and easily threw him out. For games three, four, and five in Philadelphia's National League park the designated hitter was not allowed and so in game three Gaston replaced Olerud at first with Molitor. No one could remember when a batting champion had been replaced in such a manner. The move paid off, as Toronto led by Molitor crushed the Phillies 10–3.

The pivotal and perhaps most amazing game in World Series history was played on Wednesday, 20 October in a cold drizzle. Not wishing to tempt fate twice Gaston returned Olerud to first for game four and sat Sprague in favour of Molitor at third. Toronto got off to a quick 3–0 lead but Todd Stottlemyre not only surrendered four runs in the bottom of the inning but then managed to get thrown out at third in Toronto's next at-bat, trying to go from first to third on a single to centre field. The Jays led 7–6 after three but were outscored 8–2 in the next four innings. Back in Canada even the most loyal fan, anticipating that the game would go past midnight Toronto time, got ready to retire and rest up for a tie-breaking game on Thursday. Alomar led off the eighth grounding out, but Carter doubled, Olerud walked, and Molitor also doubled. Mitch Williams entered the game and gave up hits to Fernandez and Henderson while walking Borders and striking out Sprague. The Phillies clung to the lead as Devon White strode to the plate. In 1992 his sensational catch in game three may have saved the Series for the Jays. Now his bat delivered a bases-clearing triple which skidded in the rain past the Phillie outfielders. Up 15–14 the Jays turned to Mike Timlin and Duane Ward to nail down the victory. The Jays lost the next evening 2–0 and so returned home needing one victory.

It looked to be a foregone conclusion on Saturday, 23 October as they went up 5–1, but in the seventh the Phillies scored five. In the eighth inning Toronto loaded the bases, failed to score, but in the process guaranteed that the top of the batting order would come up in the ninth. Philadelphia once again turned to Mitch Williams. Lead-off hitter Rickey Henderson got on base on a walk. White flied out but Molitor, showing his tremendous ability to make contact and get on base, lined a single to centre bringing Joe Carter to the plate. On a two and two count Carter realized his December vision from the previous year, driving an inside fastball into the left field bullpen where John Sullivan, coaching his final major league game, retrieved it. The most endearing scene other than the wildly leaping Carter was that of manager Gaston hugging Molitor who had left Milwaukee precisely because he wanted to be part of such a celebration. Amidst the celebrating crew was Dick Schofield whose father had been in a similiar tumble of humanity 33 years ago when the Pittsburgh Pirates won the World Series on Bill Mazeroski's game-winning home run. Just before the World Series Schofield had told Toronto reporter Dave Perkins that 'Maybe someone will hit a home run in the ninth inning and I'll be out there and in the picture.'

For the second year in a row Canada exploded. Toronto's Yonge Street was taken over by a generally peaceful all-night party. The next day SkyDome was filled with fans saluting their heroes, among them Jack Morris who arrived riding a policeman's motorcycle. In Saskatoon there was a small riot on 8th Street; in 1994 when the city joined the new independent North Central League they honoured that dubious moment by calling the team the Saskatoon Riot.

All great times must end sooner or later and within a year the Jays' bubble had burst. By the time of the season-ending players' strike in 1994 they found themselves 16 games out of first and about to suffer their first under-.500 season since 1982. The

Joe Carter whose game-winning home run in game six won the 1993 World Series for the Jays. Courtesy Howard Starkman, Toronto Blue Jays Baseball Club

Yankees led by the former Jay southpaw Jimmy Key compiled the best record in the American League. Over in the National the best record in all the major leagues belonged to the Montreal Expos, whose bid to make the World Series a unique Canadian triple play fell victim to the strike. Ironically, in October major league baseball declared that because of the strike the Toronto Blue Jays would enter the 1995 season as defending World Series champions.

Baseball's World Series is a historic moment in sport. Its deeds are recalled long after the memories of the final series or games in other sports have faded from remembrance. What the Jays had done in 1992 and 1993 was create moments of common reflection for Canadians which will last for several generations.

Appendix A

Canadian Big League Players

The first organized Canadian club appeared in Hamilton in 1854 and by the 1860s teams dotted southwestern Ontario. Information on the participants in general is sketchy. However, a survey of Canadian major league players may well provide a partial portrait of those who played the game in Canada in different eras.

Canadian-born players made an early appearance in major league line-ups. Among those in the first recognized league, the National Association, formed in 1871 was Guelph-born Mike Brannock, whose name reappears in this league in 1875 along with that of another Guelph-born player, Tom Smith. The Association collapsed after that year to be replaced by the National League in 1876. A year later its first serious rival, the International Association, was formed with two Canadian member teams in London and Guelph. At least eight Canadians played in the Association.

To that date not only were all of these players Ontario-born or raised, but most received their training in Canada's earliest baseball hotbeds, Guelph and London. Throughout the 1850s they were able to watch and learn from an earlier generation of pioneering ballplayers who played amateur baseball. In 1879 the first Canadian-born major leaguer from somewhere other than Ontario was Bill Phillips, born in Saint John, New Brunswick but raised from a very early age in Chicago.

The 1880s were a decade of tremendous growth in Canadian major league appearances. The peak year, 1884, was marked by 28 Canadians playing in three

major leagues, the National, American Association, and Union League. It is still the highest number in Canadian baseball history, the second highest being the 16 players in the 1993 season. The majority of ballplayers were from Ontario. They had come of baseball playing age in the early 1870s when the game was on the upward crest of its phenomenal growth in the province. Other Canadians included the first Quebecker Eugene Vadeboncoeur, and players from all four of Canada's present Atlantic provinces—'Pop' Smith from Digby, Nova Scotia; Henry Mullin from Saint John, New Brunswick; Henry Oxley, one of only two players in all of baseball history from Prince Edward Island; and the only Newfoundlander ever to play in the majors, Jim McKeever. Though many of these Canadians from outside Ontario were probably raised in the United States, the game was by the 1880s popular in at least Quebec, Nova Scotia, and New Brunswick.

For the rest of the decade the annual number of Canadians in the majors was just under 10 but ballooned to 14 in 1890 when players formed their own league to challenge the existing two. These numbers again declined through the 1890s with only one Canadian listed in 1895, followed by an annual average of five to the end of the century and then not one Canadian in 1900 (the only other years in which this had occurred were in the 1870s). At least part of the reason for this decline in the last decade of the 19th century may have been an American labour law introduced in 1894 which restricted Canadians' ability to work in the United States.

Up until the First World War Canadian major

league participation was relatively stable, returning to eight in 1902 and averaging that figure through 1905 to 1915. This period saw the entry of the American League and the establishment of baseball's present major league 16-team structure which persisted until 1961 with the brief exception of the rival Federal League in 1914–15. Major league jobs were limited and with the opening of the American Midwest and Pacific Coast to a growing American population, the opportunities for Canadian ballplayers could be expected to decline. The relative stability of their numbers in this pre-First World War era indicates the prominent role the game had assumed in Canada by the turn of the century. As well the expanding minor league system in Canada and the United States, which often offered more money and security than the big leagues, encouraged young Canadians to pursue a baseball career.

The increasing number of ballplayers, some now drawn from Cuba, and the limited number of big league jobs began to affect Canadian content almost immediately after the 1915 collapse of the Federal League. Canada entered the First World War in 1914, three years before the United States, and this also had an obvious effect on the flow of Canadians to American baseball leagues. From an average of five between 1916 through 1920, the average declined to just over two between 1921 and 1925. Canadian baseball organization suffered during the war with the result that an entire generation received less than adequate training or opportunity to play. Not surprisingly, the only Canadian to play in the majors from 1926 through 1929 was Hamilton born Frank 'Blackie' O'Rourke (b. 1912) and his significant upbringing had been in New Jersey. Nor did numbers improve in the early 1930s with the annual figure being either one or two Canadians between 1930 and 1936.

Canadian-born major leaguers were still largely from Ontario, but the Maritimes and Quebec numbers were increasing and in 1908 Bert Sincock from Barkerville, British Columbia debuted, followed a year later by Russell Ford, born in 1883 in Brandon, Manitoba. Many of these players however were only Canadian by birthplace, having learned their baseball skills from an early age in the American communities to which their families moved. Yet the very fact that Canada appeared on their birth certificate acted as a powerful incentive to aspiring Canadian ballplayers and raised Canadian spectator interest in major league games.

It speaks well for the integration of baseball into Canadian society that Canadian major league participation gained renewed strength in the late 1930s and through the 1940s, peaking at 12 players in 1946. Some of the greatest Canadian-born players of all time emerged in this era, though two in particular—George Selkirk (debut 1934) and Jeff Heath (debut 1936)—were raised from very early ages in the United States. On the other hand pitchers Phil Marchildon (debut 1940), Oscar Judd and Dick Fowler (both debut in 1941) developed their skills on Ontario playgrounds. Notably all were pitchers, a trend that dominates Canadian big league participation. Pitching, while a higher level skill than that of an outfielder, is both more one-dimensional (one did not have to have hitting skills even in the pre-designated hitter era) and allowed to evolve over a greater period of time. Scouts will tolerate a twenty-year-old pitching prospect with a great fastball and ignore a gifted twenty-year-old hitter with flaws in his swing or inferior fielding ability. Accordingly in Canada where a shorter season limits the rapid growth of baseball skills, scouts have always paid more attention to promising pitchers.

Noticeable also in the ten-year period from 1937 to 1946 were the number of ballplayers from western Canada and Quebec. Though their careers were usually short-lived their presence indicated a future source of big league talent. Numbers slowly declined in the late 1940s, though a new category of Canadian ballplayer, one born elsewhere but growing up in Canada, appeared. Hank Biasatti, born in Italy but raised in Windsor, played in 1949 and Reno Bertoia, also born in Italy and raised in Windsor, debuted for the Detroit Tigers in 1953. There were only two Canadian major leaguers that year, a number repeated again in 1956 and 1958. Curiously six out of eight major league debuts between 1959 and 1962 were by players from Quebec, including Claude Raymond (1959), Georges Maranda and Ron Piche (1960), Tim Harkness (1961), and Ray Daviault and Pete Ward (1962). The variety of their French and English Canadian origins from St-Jean, Lévis, Verdun, Lachine, and Montreal points back to their formative years in the 1940s during baseball's Golden Age in Quebec when the Provincial League and the Montreal Royals of the International League were at their peak.

The renewed Canadian representation in the majors in the 1960s is at least partly due to the expansion of children's baseball programs following the Second World

War as communities adopted this quintessential symbol of family life. Torontonian Ron Taylor for instance played in a program at Leaside's Talbot Park. When he debuted in 1962, he was one of 11 Canadians, a double digit figure again reached in 1966. Two of that year's sophomores, Ferguson Jenkins and John Hiller, had 19- and 15-year careers respectively. Canadian participation remained relatively stable through the 1970s, never dipping below five or surpassing the nine in 1977. That year marked not only the Toronto Blue Jays' inaugural American League season but the arrival in Houston of Melville, Saskatchewan's Terry Puhl who began a 15-year major league career. Through the first part of the 1980s Canadian participation actually declined to a bare two in 1984. This decline seems odd in light of the rejuvenating influence of the Montreal Expos and the Toronto Blue Jays. It accurately reflects, however, the significant decline in baseball interest in Canada in the 1960s. It would take several years to rebuild the critical mass from which future Canadian ballplayers could emerge and this renaissance of Canadian major league participation began in 1985 with players who came to baseball playing age in the 1970s. Ironically Kirk McCaskill (debut 1985) though born in Kapuskasing, Ontario has always considered himself American.

Canadian Ballplayer Inventory

Information Format

debut year (number of Canadians in majors that year) additional Canadian players in major leagues that season listed by identifying number

identifying number: (month/day of first appearance) last name, first name, nickname: (games played—seasons) birthplace (any notes on birthplace), b. birthdate, ba batting average, p pitching win-loss record

• notable comments, i.e. Hall of Fame membership

1871 (1)

1: (10/27) Brannock, Mike: (5–2), Guelph, Ont., b. 1853, ba .087
  • first Canadian in major leagues (with Chicago of the National Association in 1871 and 1875)

Numbers slowly increased through the late 1980s with the arrival of players like Rob Ducey who was 12 when the Blue Jays debuted. In 1991 there were 11 Canadian big leaguers, 13 in 1992 and, a 20th century peak, 16 in 1993. Their numbers included Rob Butler who had followed the Jays from the age of nine and realized a childhood fantasy in 1993 as a member of the World Champion Blue Jays. Nigel Wilson of the Florida Marlins—their first choice in the expansion draft—was the child of West Indies immigrants. Larry Walker from Maple River, British Columbia was a genuine born-in-Canada star for the Montreal Expos, though the fans reserved their greatest adulation for the Montreal-born French Canadian Denis Boucher whose late season arrival in the 1993 Expos line-up doubled Olympic Stadium attendance when he pitched.

That summer Boucher and Joe Siddall combined as the fifth all-Canadian pitcher/catcher battery to ever appear together in a major league game. Past batteries included pitcher James Edward 'Tip' O'Neill and John Humphries (1883), brothers Pete and Fred Wood (1885), pitcher Bob Steele and George Gibson (1918), and pitcher Ozzie Van Brabant and Eric MacKenzie (1955).

1872 (0)

1873 (0)

1874 (0)

1875 (2) 1

2: (9/15) Smith, Tom: (74–4), Guelph, Ont., b. 1851, ba .143
  • two (1877 and 1878) of four seasons in International Association

1876 (0)

1877 (8) 2

3: (/) Reid, Billy: (47–3), London, Ont., b. 1857, ba .247
  • one season (1877) in International Association
  • substitute umpire National League 1882

4:    (/) Smith, Bill: (15–1), Guelph, Ont., b. 1851, p 2–6
- listed in encyclopedias as having appeared in 1886 with Detroit but this appears to be an error
- only season in International Association

5:    (/) Maddocks, Charles: (1–1), Guelph, Ont., b.1849, ba .000
- only season in International Association
- substitute umpire National League 1882

6:    (/) Southam, Richard: (1–1), London, Ont., b. ?, ba .000
- only season in International Association
- manager of London Tecumsehs International Association team 1877
- brother of William, founder of Southam Publishing Company
- uncle of George Gibson, future major leaguer

7:    (/) Gillean, Thomas: (1–1), London, Ont., b. 1855, ba. 000
- only season in International Association
- National League umpire 1879–80, substitute 1881

8:    (/) Goldie, Thomas (information is unclear and this player may have been Thomas's brother John): (1–1), United States (listed as miller and player in Guelph as early as 1868), b. 1850, ba .500
- only season in International Association

9:    (/) Hewer, William: (1–1), England (listed as butcher and player in Guelph as early as 1869), b. ?, ba .000
- only season in International Association

1878 (1) 2

1879 (1)

10:    (5/01) Phillips, Bill: (1038–10), Saint John, N.B., b. 1857, ba .266
- member of Canadian Baseball Hall of Fame
- moved to US at young age and raised in Chicago

- hit first Canadian National League home run

1880 (3) 10

11:    (5/01) Smith, Charles 'Pop': (1093–12), Digby, N.S., b. 1856, ba .224
- career batting average of .224 is the third worst ever for players with 4,000 career at-bats

12:    (5/01) Irwin, Art 'Doc, Cutrate, Sandy': (1010–13), Toronto, Ont., b. 1858, ba .241, p 0–0
- place of death, Atlantic Ocean
- brother of major leaguer John Irwin
- umpire National League 1902
- major league manager 8 seasons
- played in 1884 post-season playoffs
- raised in Boston, Massachusetts
- member Canadian Baseball Hall of Fame
- credited with popularizing use of the fielder's glove (c. 1883) as a result of an injury to two fingers

1881 (3) 10, 11, 12

1882 (8) 2, 10, 11, 12

13:    (5/31) Irwin, John: (322–8), Toronto, Ont., b. 1861, ba .246
- brother of major leaguer Art Irwin
- played in 4 different major leagues (National, American Association, Union, Players)

14:    (7/17) Casey, Orrin Robinson 'Bob': (9–1), Adolphustown, Ont., b. 1859, ba .231

15:    (7/26) Doyle, John: (3–1), Nova Scotia, b. 1858, p 0–3
- raised in Providence, Rhode Island

16:    (8/31) Thompson, John 'Tug': (25–2), London, Ont., b. ?, ba .206

1883 (9) 3, 10, 11, 12

17:    (5/05) O'Neill, James Edward 'Tip':

(1054–10), Woodstock, Ont., b. 1858, ba
.326, p 16–16
  • batted .435 in 1887 with St Louis of
    American Association (.490 on base per-
    centage, .691 slugging average)
  • played in 1885, 1886, 1887, and 1888
    post-season playoffs
  • later part owner of Eastern League
    Montreal Royals
  • member of Canada's Sports Hall of Fame
    and the Canadian Baseball Hall of Fame
  • career batting average of .326, 30th high-
    est career batting average in baseball his-
    tory among regular players

18: (7/07) Humphries, John: (98–2), North
     Gower, Ont., b. 1861, ba .143

19: (7/25) Emslie, Bob: (91–3), Guelph, Ont., b.
     1859, p 44–44
       • 32–17 in 1884 with Baltimore of
         American Association
       • umpire American Association 1890,
         National League 1891–1924
       • on Baseball Hall of Fame Honour Roll
       • member Canadian Baseball Hall of Fame

20: (9/25) Pirie, James Moir: (5–1), Dundas,
     Ont., b. ?, ba .158
       • stats listed incorrectly under Dick Pierre
         in encyclopedias

21: (9/29) Mountjoy, Bill 'Medicine Bill': (57–3),
     London, Ont., b. 1857, p 31–24
       • encyclopedias incorrectly show birth-
         place as Port Huron, Michigan
       • death year 1894 not 1934 as shown in
         encyclopedias
       • 19–12 in 1884 with Cincinnati of
         American Association

1884 (28) 3, 10, 11, 12, 13, 16, 17, 18, 19, 21

22: (4/17) McKeever, Jim: (16–1), St John's,
     Newfoundland (recent research indicates
     McKeever may have been born in Saint John,
     N.B.), b. 1861, ba .136

23: (4/20) Whitehead, Milt: (104–1), Canada, b.
     1862, ba .207, p 0–1

24: (5/01) Collins, Charles 'Chub': (97–2),
     Dundas, Ont., b. 1857, ba .196

25: (5/02) Hunter, Bill: (2–1), St Thomas, Ont.,
     b. ?, ba .143

26: (5/02) Knowles, Jimmy 'Darby': (357–5),
     Toronto, Ont., b. 1859, ba .241

27: (5/10) Gardner, Alex: (1–1), Toronto, Ont.,
     b. 1861, ba .000

28   (5/14) Wood, Fred: (13–2), Hamilton, Ont.,
     b. 1863, ba .065

29: (5/16) Knight, Jonas 'Quiet Joe': (133–2),
     Port Stanley, Ont., b. 1859, ba .309, p 2–4

30: (6/04) Mullin, Henry: (36–1), Saint John,
     N.B., b. 1862, ba .133

31: (7/04) Scanlan, Patrick: (6–1), N.S., b. 1861, ba .292

32: (7/09) Dorsey, Jerry: (1–1), Canada, b. 1854,
     ba .000, p 0–1

[Art Richardson from Hamilton, Ont. is listed in the
baseball encyclopedias as having played a major league
game on July 10, 1884. Subsequent research however
has shown that the Art Richardson who played that
game was a Boston amateur. The Art Richardson from
Hamilton was playing for Oil City at the time.]

33: (7/11) Vadeboncoeur, Onesime Eugene:
     (4–1), Louiseville, Que., b. 1858, ba .214

34: (5/30) Weber, Joe: (2–1), Hamilton, Ont., b.
     1861, ba .000
       • encyclopedias incorrectly show Joe
         Weber with Indianapolis in 1884. Joe
         Weber's correct stats with Detroit in 1884
         are shown as those of Harry Weber

35: (7/30) Oxley, Henry: (4–1), Covehead, P.E.I.,
     b. 1858, ba .000
       • moved to Massachusetts at age of 4

36: (8/01) Watkins, Bill: (34–1), Brantford, Ont.,
     b. 1858, ba .205
       • major league manager 9 seasons

37:   (8/01) Morrison, Jon: (53–2), London, Ont.,
      b. 1859, ba .241
        • encyclopedias incorrectly show Port
          Huron, Michigan as place of birth

38:   (8/06) Smith, Frank: (10–1), Canada, b.
      1857, ba .250

39:   (9/27) Dunn, Steve: (9–1), London, Ont., b.
      1858, ba .250

1885 (10) 10, 11, 12, 17, 19, 21, 24, 28

40:   (7/15) Wood, Pete: (27–2), Hamilton, Ont.,
      b. 1857, p 9–16
        • brother of major leaguer Fred Wood
          (teammates with Buffalo of the
          National League in 1885)

41:   (9/15) Andrus, Wyman: (1–1), Orono, Ont.,
      b. 1858, ba .000
        • University of Toronto graduate and
          mayor of Miles City, Montana

1886 (7) 10, 11, 12, 13, 17, 26

42:   (7/23) Hyndman, Jim: (1–1), province of
      Ontario, b. 1865, ba .000, p 0–1

1887 (8) 10, 11, 12, 13, 17, 26, 37

43:   (5/03) O'Neill, Fred 'Tip': (6–1), London,
      Ont., b. 1865, ba .308

1888 (7) 10, 11, 12, 13, 17

44:   (7/30) LaRoque, Sam: (124–3), St Mathias,
      Que., b. 1864, ba .249

45:   (8/01) Walker, George: (4–1), Hamilton,
      Ont., b. 1863, p 1–3

1889 (6) 11, 12, 13, 17, 40

46:   (4/18) Johnson, John 'Spud': (331–3),
      Canada, b. 1860, ba .302
        • batted .346 in 135 games in 1890 for
          Columbus of American Association

1890 (14) 11, 12, 13, 17, 26, 29, 44, 46

47:   (6/03) O'Connor, Dan: (6–1), Guelph, Ont.,
      b. 1868, ba .462

48:   (7/14) Osborne, Fred: (41–1), Hampton,
      Iowa (listed as such in encyclopedias but ear-
      lier records indicate canada), b. ?, ba .238, p
      0–5

49:   (7/21) Lyons, Pat: (11–1), Canada, b. 1860,
      ba .053

50:   (7/26) Demarais, Fred: (1–1), province of
      Quebec (some sources say Nashua, New
      Hampshire), b. 1866, p 0–0

51:   (8/12) Jones, Mike: (3–1), Hamilton, Ont., b.
      ?, p 2–0

52:   (10/01) Gillespie, Jim: (1–1), Canada, b.
      1858, ba .000

1891 (8) 11, 12, 13, 17, 44, 46

53:   (4/22) O'Brien, John 'Chewing Gum':
      (501–6), Saint John, N.B., b. 1870, ba .254

54:   (5/07) Lake, Fred: (48–5), province of Nova
      Scotia, b. 1866, ba .232
        • played with Boston of National League in
          1897 Temple Cup
        • major league manager 3 seasons
        • 11-season gap (1898 to 1910) between
          major league player appearances

1892 (2) 17, 26

1893 (3) 53

55:   (6/12) Yost, Gus: (1–1), ? (birthplace
      unknown though earlier encyclopedias show
      Toronto, Ontario), b. ?, p 0–1

56:   (8/05) Summers, William 'Kid': (2–1),
      Toronto, Ont., b. ?, ba .000

1894 (3) 12, 54

57:   (6/16) Pfann, Bill: (1–1), ? (encyclopedias list
      a Bill Pflann born in Brooklyn, but according
      to researcher Bill Carle the Brooklyn city

directory of the period lists a Bill Pfann, a player with Hamilton in 1886, as being born in Canada, probably in Hamilton), b. ?, p 0–1

1895 (1) 53

1896 (4) 53

58: (4/18) Payne, Harley 'Lady': (80–4), Windsor, Ont., b. 1868, p 30–36

59: (5/02) Hulen, Billy: (107–2), Dixon, California (early in his career when he was playing for Detroit of the International League, the Toronto Globe said that he was born in Whitby, Ontario—historian Eves Raja speculates they got the information from Hulen), b. 1870, ba .246

60: (9/01) Johnson, Albert: (73–2), London, Ont., b. 1872, ba .238
  • incorrectly listed in encyclopedias as pitcher Abe Johnson; his statistics appear in encyclopedias as those of Swedish-born Abbie Johnson

1897 (6) 53, 54, 58, 60

61: (4/29) Hannifin, Pat: (10–1), province of Nova Scotia, b. 1868, ba .250

62: (5/18) Magee, Bill: (106–5), Canada, b. 1875, p 29–51
  • pitched 295 innings with Louisville of National League in 1898

1898 (4) 54, 58, 62

63: 1(5/19) Snyder, Frank 'Cooney': (17–1), Toronto, Ont., b. ?, ba .164

1899 (5) 53, 58, 59, 62

64: (9/02) Frisk, Emil: (158–4), Kalkaska, Michigan (1899 issue of *Hamilton Spectator* says Frisk was born in northern Ontario town of Ignace), b. 1874, ba .267, p 8–10

1900 (0)

1901 (4) 62, 64

65: (4/26) McLean, John B. 'Larry': (862–13), (baseball encyclopedias indicate Cambridge, Massachusetts but according to reporter Doug Black in March 1921, 'John B. McLean was born here in Fredericton in 1881. When he was a wee kiddie his parents moved to Boston, but some of his relatives still reside here. Larry drifted back here in the latter 90s . . .'), b. 1881, ba .262

66: (4/26) Kellum, Win: (48–3), Waterford, Ont., b. 1876, p 20–16
  • 15–10 record with Cincinnati of National League in 1904

1902 (8) 62

67: (4/18) Congalton, William 'Bunk': (307–4), Guelph, Ont., b. 1875, ba .290
  • hit .320 in 117 games with Cleveland of American League in 1906

68: (4/25) Currie, Clarence: (53–2), Glencoe, Ont., b. 1878, p 15–23

69: (7/21) LePine, Louis 'Pete': (30–1), Montreal, Que., b. 1876, .208

70: (9/04) Hardy, Alex 'Dooney' (7–2), Toronto, Ont., b. 1877, p 3–3

71: (9/11) Long, Nelson 'Red': (1–1), Burlington, Ont., b. 1876, p 0–0

72: (9/17) Ross, Ernie 'Curly': (2–1), Toronto, Ont., b. 1880, p 1–1

73: (9/21) Vickers, Harry 'Rube': (88–5), St Marys, Ont., b. 1878, p 22–27
  • 18–19 pitching record for Philadelphia of American League in 1908 (317 innings pitched)
  • set organized baseball record pitching 517 innings, striking out 409 batters in 64 starts in 1906 for Seattle of the Pacific Coast League
  • after winning the first game of a double-header on October 5, 1907, Vickers

pitched a five-inning perfect game in the second

1903 (5) 65, 68, 70, 73

74:　(9/14) Pinnance, Ed 'Peanuts': (2–1), Walpole Island, Ont., b. 1879, p 0–0

1904 (4) 65, 66

75:　(5/07) O'Neill, Bill: (206–2), Saint John, N.B., b. 1880, ba .243

76:　(9/06) Archer, Jimmy: (847–12), Dublin, Ireland (raised from infancy in Toronto, Ontario), b. 1883, ba .249
* played in World Series with Detroit of American League in 1907 and Chicago of National League in 1910
* member Canadian Baseball Hall of Fame

1905 (10) 64, 66, 67

77:　(4/21) McGovern, Art: (15–1), Saint John, N.B., b. 1882, ba .114

78:　(4/25) Hogg, William 'Buffalo Bill': (116–4), Canada (encyclopedias show Port Huron, Michigan but according to historian Peter Morris recent records indicate Canada), b. 1879, p 37–50
* died in New Orleans in 1909 after being hit by a pitch

79:　(4/26) Clarke, Jay 'Nig': (506–9), Amherstburg, Ont., b. 1882, ba .254
* hit 8 home runs in one minor league game in Texas
* credited by Ty Cobb as being the first catcher to wear shin guards
* catcher for Addie Joss's perfect game

80:　(5/05) Ford, Gene: (7–1), Milton, N.S., b. 1881, p 0–1
* brother of major leaguer Russ Ford
* died in Blue Jays' spring training home, Dunedin, Florida

81:　(7/02) Gibson, George 'Moon': (1213–14), London, Ont., b. 1880, ba .236

* played for Pittsburgh of National League in 1909 World Series
* major league manager 7 seasons, major league coach 2 seasons
* member of Canada's Sports Hall of Fame
* member Canadian Baseball Hall of Fame

82:　(9/11) Owens, Frank 'Yip': (222–4), Toronto, Ont., b. 1886, ba .245
* debuted in majors with Boston of American League at age of 19

83:　(9/28) Cockman, Jim: (13–1), Guelph, Ont., b. 1873, ba .105

1906 (7) 65, 67, 75, 78, 79, 81

84:　(9/13) Cameron, John 'Happy Jack': (18–1) province of Nova Scotia, b. 1884, ba .180, p 0–0

1907 (9) 64, 65, 67, 73, 76, 78, 79, 81

85:　(4/18) Randall, Newton 'Newt': (97–1), New Lowell, Ont., b. 1880, ba .211

1908 (7) 65, 73, 78, 79, 81

86:　(4/30) Graney, Jack: (1402–14), St Thomas, Ont., b. 1886, ba .250, p 0–0
* first major league batter to face the rookie pitcher, Babe Ruth
* played for Cleveland of American League in 1920 World Series
* first major leaguer to be radio play-by-play broadcaster (Cleveland 1932–53)
* member Canadian Baseball Hall of Fame

87:　(6/25) Sincock, Bert: (1–1), Barkerville, B.C., b. 1887, p 0–0

1909 (8) 65, 73, 76, 79, 81, 82

88:　(4/15) O'Hara, Bill: (124–2), Toronto, Ont., b. 1883, ba .232, p 0–0

89:　(4/28) Ford, Russ: (199–7), Brandon, Man., b. 1883, p 99–71
* 3 twenty-plus win seasons
* member Canadian Baseball Hall of Fame

1910 (9) 54, 65, 76, 79, 81, 86, 88, 89

90:   (5/04) Miller, Roy 'Doc': (557–5), Chatham, Ont., b. 1883, ba .295
- in 1911 with Boston of National League hit .333 (second in league to Honus Wagner) with 192 hits (first in league)
- third best pinch hitting average in history— .325

1911 (9) 65, 76, 79, 81, 86, 89, 90

91:   (5/27) Rowan, Dave: (18–1), Elora, Ont., b. 1882, ba .385
- name at birth, Dave Drohan
- career batting average of .385 highest ever for batters with more than 50 at-bats

92:   (6/20) Jones, Bill 'Midget': (27–2), Hartland, N.B., b. 1887, ba .226

1912 (9) 65, 76, 81, 86, 89, 90, 92

93:   (6/12) O'Rourke, Frank 'Blackie': (1131–14), Hamilton, Ont., b. 1894, ba .254
- played 14 seasons over a 20-year span
- family moved to US in 1899; O'Rourke raised in New Jersey
- scout for Cincinnati of National League 1941–51, and with Yankees 1952–83

94:   (9/07) Kyle, Andy: (9–1), Toronto, Ont., b. 1889, ba .333

1913 (7) 65, 76, 81, 86, 89, 90

95:   (9/23) Daly, Tom: (244–8), Saint John, N.B., b. 1891, ba .239
- coach with Boston of American League for 14 seasons (1933–46)

1914 (8) 65, 76, 81, 82, 86, 89, 90, 95

1915 (8) 65, 76, 81, 82, 86, 89, 95

96:   (9/14) Dee, Maurice 'Shorty': (1–1), Halifax, N.S., b. 1889, ba .000
- moved to US at young age and grew up in Boston, Mass.

1916 (5) 76, 81, 86, 95

97:   (4/17) Steele, Bob: (91–4), Cassburn, Ont., b. 1894, p 16–38

1917 (5) 76, 81, 86, 93, 97

1918 (6) 76, 81, 86, 93, 95, 97

1919 (4) 79, 86, 95, 97

1920 (5) 79, 86, 93, 95

98:   (10/02) Wingo, Ed: (1–1), Ste Anne de Bellevue, Que., b. 1895, ba .250
- name at birth, Edmond Armand La Rivière

1921 (4) 86, 93, 95

99:   (7/03) Riley, Jim: (6–2), Bayfield, N.B., b. 1895, ba .000

1922 (3) 86, 93

100:   (9/15) O'Neill, Harry: (4–2), Ridgetown, Ont., b. 1897, p 0–0
- University of Toronto grad and one of first to throw the slider

1923 (2) 99, 100

1924 (2) 93

101:   (9/20) Shields, Vince: (2–1), Fredericton, N.B., b. 1900, p 1–1

1925 (2) 93

102:   (9/16) Kerr, Mel: (1–1), Souris, Man., b. 1903, no batting average
- only major league appearance as a pinch runner for Chicago of the National League on September 16, 1925

1926 (1) 93

1927 (1) 93

1928 (1) 93

1929 (1) 93

1930 (2) 93

103: (9/17) Dugas, Augustin (Gus): (125–4), St Jean Dematha, Que., b. 1907, ba .206

1931 (2) 93

104: (7/17) Barton, Vince: (102–2), Edmonton, Alta., b. 1908, ba .233

1932 (2) 103, 104

1933 (1) 103

1934 (2) 103

105: (8/12) Selkirk, George 'Twinkletoes': (846–9), Huntsville, Ont., b. 1908, ba .290
- appeared in six World Series with the New York Yankees
- appeared in two All-Star games, 1936 and 1939
- replaced Babe Ruth in the Yankee outfield
- family moved to Rochester, New York when Selkirk was three
- 108 career home runs, 5 seasons with plus .300 batting average
- member of Canadian Baseball Hall of Fame

1935 (1) 105

1936 (2) 105

106: (9/13) Heath, Jeff: (1383–14) Fort William, Ont., b. 1915, ba .293
- first American Leaguer 20-plus doubles, triple and homers in a season (1941— 32/20/24)
- 7 seasons with plus .300 batting average including 1938 (.343 in 502 at bats) and 1941 (.340 in 585 at-bats)
- most career hits by a Canadian, 1,447, and career home runs, 194
- appeared in All-Star game in 1941 and 1943 for Cleveland of American League
- raised in Seattle, Washington
- member of Canadian Baseball Hall of Fame

1937 (4) 105, 106

107: (9/09) Krakauskas, Joe: (149–7), Montreal, Que., b. 1915, p 26–36

108: (9/14) Rosen, Goodwin 'Goody': (551–6), Toronto, Ont., b. 1912, ba .291
- in 1945 hit .325 for Brooklyn Dodgers and named to National League All-Star team
- member of Canadian Baseball Hall of Fame

1938 (5) 105, 106, 107, 108

109: (9/11) Buxton, Ralph 'Buck': (19–2), Weyburn, Sask., b. 1911, p 0–2
- 10-season gap between major league appearances (1938–49)

1939 (4) 105, 106, 107, 108

1940 (5) 105, 106, 107

110: (9/08) Robertson, Sherry: (597–10), Montreal, Que., b. 1919, ba .230
- brother of Washington Senators/ Minnesota Twins owner Calvin Griffith
- Farm Director of Washington Senators 1956–70

111: (9/22) Marchildon, Phil 'Babe': (185–9), Penetang, Ont., b. 1913, p 68–75
- won 17 games in 1942 and 19 in 1947, for Philadelphia of American League
- against Marchildon, DiMaggio hit safely in game 46 of his 56-game consecutive hitting streak in 1941
- member of Canada's Sports Hall of Fame and Canadian Baseball Hall of Fame

1941 (9) 105, 106, 107, 110, 111

112: (4/16) Judd, Oscar 'Ossie': (161–8) London, Ont., b. 1908, p 40–51
- member of Canadian Baseball Hall of Fame

113: (4/22) Wilkie, Aldon 'Lefty': (68–3), Zealandia, Sask., b. 1914, p 8–11
- member Saskatchewan Baseball Hall of Fame

114: (9/12) Cook, Earl: (1–1), Stouffville, Ont., b. 1908, p 0–0

115: (9/13) Fowler, Dick: (221–10) Toronto, Ont., b. 1921, p 66–79
   • on September 9, 1945 as member of Philadelphia Athletics became only Canadian to throw a no-hitter
   • member of Canadian Baseball Hall of Fame

1942 (10) 105, 106, 107, 111, 112, 113, 115

116: (4/14) Sketchley, Harry 'Bud': (13–1), Virden, Man., b. 1919, ba .194

117: (9/12) Colman, Frank: (271–6), London, Ont., b. 1918, ba .228

118: (9/24) Calvert, Paul 'Cyclops': (109–7), Montreal, Que., b. 1917, p 9–22

1943 (6) 106, 110, 112, 117, 118

119: (8/28) Mead, Charlie: (87–3), Vermilion, Alta., b. 1921, ba .245

1944 (7) 106, 108, 112, 117, 118, 119

120: (4/18) Gladu, Roland: (21–1), Montreal, Que., b. 1913, ba .242

1945 (9) 106, 108, 111, 112, 115, 117, 118, 119

121: (8/04) LaForest, Byron 'Ty': (52–1), Edmundston, N.B., b. 1917, ba .250

1946 (12) 106, 107, 108, 110, 111, 112, 113, 115, 117

122: (5/01) Bahr, Ed: (46–2), Rouleau, Sask., b. 1919, p 11–11

123: (5/05) Roy, Jean-Pierre: (3–1), Montreal, Que., b. 1920, p 0–0

124: (9/18) McCabe, Ralph 'Mack': (1–1), Napanee, Ont., b. 1918, p 0–1
   • surrendered home run to first major league batter he faced (another Canadian, Sherry Robertson)

1947 (7) 106, 110, 111, 112, 115, 117, 122

1948 (5) 106, 110, 111, 112, 115

1949 (7) 106, 109, 110, 111, 115, 118

125: (4/23) Biasatti, Hank: (21–1), Beano, Italy (raised in Windsor, Ontario), b. 1922, ba .083
   • member of Toronto Huskies basketball team 1946–47

1950 (6) 110, 111, 115, 118

126: (4/19) Hooper, Bob: (194–6), Leamington, Ont., b. 1922, p 40–41
   • 15–10 for Philadelphia Athletics in rookie season 1950

127: (5/09) Erautt, Joe 'Stubby': (32–2), Vibank, Sask., b. 1921, ba .186
   • brother (also a major leaguer) born three years later in Oregon, suggesting family moved to United States

1951 (6) 110, 115, 118, 126, 127

128: (9/16) Fisher, Harry: (18–2), Newbury, Ont., b. 1926, ba .278, p 1–2

1952 (5) 110, 115, 126, 128

129: (4/30) Rutherford, Johnny 'Doc': (22–1), Belleville, Ont., b. 1925, p 7–7
   • played in 1952 World Series with Brooklyn Dodgers

1953 (2) 126

130: (9/22) Bertoia, Reno: (612–10), St Vito Udine, Italy (father a Canadian citizen, Reno raised in Windsor, Ontario), b. 1935, ba .244
   • debuted in major leagues at age of 18 with Detroit Tigers
   • member of Canadian Baseball Hall of Fame

1954 (4) 126, 130

131: (4/13) Van Brabant, Camille Oscar 'Ozzie':

(11–2), Kingsville, Ont., b. 1926, p 0–2
  • moved to United States at age of three

132: (4/17) Burgess, Tom 'Tim': (104–2), London, Ont., b. 1927, ba .177
  • member Canadian Baseball Hall of Fame

1955 (6) 126, 130, 131

133: (4/11) Alexander, Bob: (9–2), Vancouver, B.C., b. 1922, p 1–1

134: (4/11) Gorbous, Glen: (117–3), Drumheller, Alta., b. 1930, ba .238
  • on August 1, 1957 at a field day in Omaha, Nebraska, Gorbous threw a ball a world record distance of 445 feet, 3 inches

135: (4/23) MacKenzie, Eric: (1–1), Glendon, Alta., b. 1932, ba .000

1956 (2) 130, 134

1957 (4) 130, 133, 134

136: (9/27) Harris, Bill: (2–2), Duguayville, N.B., b. 1931, p 0–1

1958 (2) 130

137: (7/20) Bowsfield, Ted: (215–7), Vernon, B.C., b. 1935, p 37–39
  • member Canadian Baseball Hall of Fame

1959 (4) 130, 136, 137

138: (4/15) Raymond, Claude 'Frenchy': (449–12), St Jean, Que., b. 1937, p 46–53
  • 83 major league saves including 23 for the Montreal Expos in 1970
  • member Canadian Baseball Hall of Fame

1960 (5) 130, 137

139: (4/26) Maranda, Georges: (49–2), Lévis, Que., b. 1932, p 2–7

140: (5/02) MacKenzie, Ken: (129–6), Gore Bay, Ont., b. 1934, p 8–10

141: (5/30) Piche, Ron: (134–6), Verdun, Que., b. 1935, p 10–16
  • coach Montreal Expos 1976
  • member Canadian Baseball Hall of Fame

1961 (6) 130, 137, 138, 140, 141

142: (9/12) Harkness, Tim: (259–4), Lachine, Que., b. 1937, ba .235
  • tied National League record for latest grand slam in a game (14 inning) for New York Mets in 1963
  • hit first base hit in history of Shea Stadium on April 17, 1964

1962 (11) 130, 132, 137, 138, 139, 140, 141, 142

143: (4/11) Taylor, Ron: (491–11), Toronto, Ont., b. 1937, p 45–43
  • 72 career major league saves
  • pitched in World Series for St Louis Cardinals (1964) and New York Mets (1969)
  • graduate of North Toronto Collegiate and University of Toronto
  • member Canadian Baseball Hall of Fame
  • member Canada's Sports Hall of Fame

144: (4/13) Daviault, Ray: (36–1), Montreal, Que., b. 1934, p 1–5

145: (9/21) Ward, Pete: (973–9), Montreal, Que., b. 1939, ba .254
  • in 1963, 22 homers and .295 ba for Chicago White Sox, and in 1964, 23 homers and .282 ba
  • coach Atlanta Braves 1978
  • member Canadian Baseball Hall of Fame

1963 (8) 137, 138, 140, 141, 142, 143, 145

146: (5/30) Lawrence, Jim: (2–1), Hamilton, Ont., b. 1939, no batting average (played in defensive role)

1964 (8) 137, 138, 140, 142, 143, 145

147: (4/14) Handrahan, Vern: (34–2), Charlottetown, P.E.I., b. 1938, p 0–2

148: (9/06) Craig, Pete: (6–3), Lasalle, Ont., b. 1940, p 0–3

1965 (9) 138, 140, 141, 143, 145, 148

149: (5/07) Harrison, Tom: (2–1), Trail, B.C., b. 1945, p 0–0

150: (9/06) Hiller, John: (545–15), Toronto, Ont., b. 1943, p 87–76
  • World Series (1968) and post-season playoffs (1972) with Detroit Tigers
  • All-Star (1974) for Detroit with whom he spent entire career
  • 125 career saves including 38 in 1973
  • member Canadian Baseball Hall of Fame
  • 38 saves in 1973 for Detroit Tigers, most for a lefthander until 1986

151: (9/10) Jenkins, Ferguson 'Fergie': (664–19), Chatham, Ont., b. 1943, p 284–226
  • seven 20-plus win seasons including six consecutive with the Chicago Cubs (1967–72)
  • three All-Star elections (1967, 1971, 1972), and Cy Young Award in National League (1971)
  • five seasons 300-plus innings pitched including 325 innings (1971), and 328 innings (1974)
  • strikeout-to-walk ratio better than 3:1 (3192–997)
  • member Canadian Baseball Hall of Fame
  • member Canada's Sports Hall of Fame
  • only Canadian member of National Baseball Hall of Fame in Cooperstown, New York

1966 (10) 138, 141, 143, 145, 147, 148, 150, 151

152: (4/16) Lines, Dick: (107–2), Montreal, Que., b. 1938, p 7–7

153: (9/10) Korince, George 'Moose': (11–2), Ottawa, Ont., b. 1946, p 1–0

1967 (8) 138, 143, 145, 150, 151, 152, 153

154: (4/16) Upham, John: (21–2), Windsor, Ont., b. 1941, ba .308, p 0–1

1968 (6) 138, 143, 145, 150, 151, 154

1969 (8) 138, 143, 145, 150, 151

155: (4/11) Kilkenny, Mike: (139–5), Bradford, Ont., b. 1945, p 23–18
  • played for 4 teams in 1972 (Detroit, Oakland, San Diego, Cleveland)

156: (6/29) Law, Ron: (35–1), Hamilton, Ont., b. 1946, p 3–4

157: (10/01) Cleveland, Reggie 'Snacks': (428–13), Swift Current, Sask., b. 1948, p 105–106
  • pitched for Boston Red Sox in 1975 World Series
  • member of Saskatchewan Baseball Hall of Fame
  • member Canadian Baseball Hall of Fame

1970 (8) 138, 143, 145, 150, 151, 155, 157

158: (5/16) Shank, Harvey: (1–1), Toronto, Ont., b. 1946, p 0–0

1971 (5) 138, 143, 151, 155, 157

1972 (5) 143, 150, 151, 155, 157

1973 (6) 150, 151, 155, 157

159: (7/01) Pagan, Dave: (85–5), Nipawin, Sask., b. 1949, p 4–9
  • member Saskatchewan Baseball Hall of Fame

160: (8/05) Ostrosser, Brian: (4–1), Hamilton, Ont., b. 1949, ba .000

1974 (5) 150, 151, 157, 159

161: (9/10) Balaz, John: (59–2), Toronto, Ont., b. 1950, ba .241

1975 (7) 150, 151, 157, 159, 161

162: (8/05) Crosby, Ken: (16–2), New Denver, B.C., b. 1947, p 1–0

163: (8/22) McKay, Dave: (645–8), Vancouver, B.C., b. 1950, ba .229

- coach Oakland Athletics 1984–94
- first Canadian to play for Toronto Blue Jays

1976 (8) 150, 151, 157, 159, 162, 163

164: (9/16) Landreth, Larry: (7–2), Stratford, Ont., b. 1955, p. 1–4

165: (9/18) Atkinson, Bill: (98–4), Chatham, Ont., b. 1954, p 11–4

1977 (9) 150, 151, 157, 159, 163, 164, 165

166: (4/22) Cort, Barry: (7–1), Toronto, Ont., b. 1956, p 1–1

167: (7/12) Puhl, Terry: (1531–15), Melville, Sask., b. 1956, ba .280
   - six seasons of perfect fielding
   - National League All-Star 1978 for Houston Astros
   - played in post-season with Houston in 1980, 1981, and 1986
   - hit a championship series record .526 in 1980
   - three plus .300 ba seasons
   - member Saskatchewan Baseball Hall of Fame
   - career .993 fielding average (19 errors in 2,596 chances) best for an outfielder
   - most major league games played by a Canadian, 1,531
   - most career stolen bases by a Canadian, 217

1978 (7) 150, 151, 157, 163, 165, 167

168: (9/04) Burnside, Sheldon: (19–3), South Bend, Indiana (raised in Toronto, Ontario), b. 1954, p 2–1

1979 (8) 150, 151, 157, 163, 165, 167, 168

169: (9/07) Pladson, Gordie: (20–4), New Westminster, B.C., b. 1956, p 0–4

1980 (8) 150, 151, 157, 163, 167, 168, 169

170: (8/31) Hodgson, Paul: (20–1), Montreal, Que., b. 1960, ba .220

1981 (6) 151, 157, 163, 167, 169

171: (5/09) Lisi, Rick: (9–1), Halifax, N.S., b. 1956, ba .313

1982 (5) 151, 163, 167, 169

172: (9/05) Frobel, Doug: (268–5), Ottawa, Ont., b. 1959, ba .201

1983 (3) 151, 167, 172

1984 (2) 167, 172

1985 (4) 167, 172

173: (5/01) McCaskill, Kirk: (296–10), Kapuskasing, Ont., b. 1961, p 95–99
   - played in post-season with California Angels (1986), and Chicago White Sox (1993)
   - son of a Canadian professional hockey player
   - lived in Kapuskasing two years before family moved to United States

174: (8/09) Shipanoff, Dave: (26–1), Edmonton, Alta., b. 1959, p 1–2

1986 (2) 167, 173

1987 (4) 167, 172, 173

175: (5/01) Ducey, Rob: (251–8), Toronto, Ont., b. 1965, ba .239
   - graduate of Ontario's Inter-County Major Baseball League (others who went on to major leagues include Jesse Orosco and Geoff Zahn)

1988 (5) 167, 173, 175

176: (9/13) Reimer, Kevin: (488–6), Macon, Georgia (father a Canadian citizen, Kevin raised in Enderby, British Columbia), b. 1964, ba .258
   - father a minor league outfielder
   - member of 1984 Canadian Olympic baseball team

177: (9/16) Wilson, Steve 'Slapshot': (205–6), Victoria, B.C., b. 1964, p 13–18
- member of 1984 Canadian Olympic baseball team

1989 (6) 167, 173, 175, 176, 177

178: (8/16) Walker, Larry: (674–6), Maple River, B.C., b. 1966, ba .281
- National League All-Star 1992–93 for Montreal Expos

1990 (7) 167, 173, 175, 176, 177, 178

179: (9/08) Gardiner, Mike: (117–5), Sarnia, Ont., b. 1965, p 17–27
- member of 1984 Canadian Olympic baseball team

1991 (11) 167, 173, 175, 176, 177, 178, 179

180: (4/12) Boucher, Denis: (35–4), Montreal, Que., b. 1968, p 6–11

181: (8/03) Wainhouse, Dave: (5–2), Toronto, Ont., b. 1967, p 0–1
- raised in Seattle, Washington
- member of 1988 Canadian Olympic baseball team

182: (8/15) Cormier, Rheal: (87–4), Moncton, N.B., b. 1967, p 24–23
- member of 1988 Canadian Olympic baseball team

183: (9/05) Horsman, Vince: (135–4), Halifax, N.S., b. 1967, p 4–2

1992 (13) 173, 175, 176, 177, 178, 179, 180, 182, 183

184: (4/11) Hoy, Peter: (5–1), Brockville, Ont., b. 1966, p 0–0

185: (5/29) Stairs, Matt: (19–2), Saint John, N.B., b. 1969, ba .210

186: (7/08) Maysey, Matt: (25–2), Hamilton, Ont., b. 1967, p 1–2
- in 1993 became one of eight American League pitchers, since introduction of designated hitter rule in 1973, to get a base hit (Ferguson Jenkins is another)

187: (7/20) Quantrill, Paul: (111–3), London, Ont., b. 1968, p 11–18

1993 (16) 173, 175, 176, 177, 178, 179, 180, 181, 182, 183, 185, 186, 187

188: (6/12) Butler, Rob: (35–2), Toronto, Ont., b. 1970, ba .213
- member of 1988 Canadian Olympic baseball team
- first Canadian to play on a Canadian World Series champion (Toronto 1993)

189: (7/28) Siddall, Joe: (19–1), Windsor, Ont., b. 1967, ba .100

190: (9/08) Wilson, Nigel: (7–1), Oshawa, Ont., b. 1970, ba .000
- first player drafted by new Florida Marlins National League franchise

1994 (11) 173, 175, 178, 179, 180, 182, 183, 187, 188

191: (5/06) Spoljaric, Paul: (2–1), Kelowna, B.C., b. 1970, p 0–1

192: (5/16) Greg O'Halloran: (12–1), Mississauga, Ont., b. 1968, ba .182
- member of 1988 Canadian Olympic baseball team

Appendix B

Canadian Women in All American Girls Professional Baseball League 1943–54

Player	Home Town	Career Data
Velma Abbott	Regina, Saskatchewan	1946–48
Janet Anderson (Perkin)	Bethune, Saskatchewan	1946
Mary (George) Baker 'Bonnie'	Regina, Saskatchewan	1943–1950, 1952 b. 1919 921 games, career ba .235; player/manager, Kalamazoo, 1950
Barbara Barbaze	Toronto, Ontario	1948
Doris Barr	Winnipeg, Manitoba	1944–46
Catherine Bennett	Regina, Saskatchewan	1943–44
Betty Berthiaume (Wickin)	Regina, Saskatchewan	1945–46
Helen Callaghan (St Aubin)	Vancouver, British Columbia	1944–46, 1948 b. 1923 388 games, 354 stolen bases, career ba .257, won 1945 batting crown (ba .299); son Casey Candaele, a major league ballplayer
Margaret Callaghan (Maxwell)	Vancouver, British Columbia	1944–51 b. 1921 672 games, career ba .196; most putouts at 3rd base, 1946
Eleanor Callow	Winnipeg, Manitoba	1948–54
Virginia Carrigy	Regina, Saskatchewan	1944
Muriel Coben	Saskatoon, Saskatchewan	1943
Dorothy Cook	St Catharines, Ontario	1946
Gladys Davis 'Terrie'	Toronto, Ontario	1943–47 batting champ in 1943, .332

Terry Donahue	Melaval, Saskatchewan	1946–49
Julianna Dusanko	Regina, Saskatchewan	1944
June Emerson	Saskatchewan	1948–49
Dorothy Ferguson (Key)	Winnipeg, Manitoba	1945–54
Gene George (McFaul)	Regina, Saskatchewan	1948
Thelma Grambo (Hundeby)	Domremy, Saskatchewan	1946
Olga Grant	Calgary, Alberta	1944
Audrey Haines (Daniels)	Winnipeg, Manitoba	1944–48, 1951 b. 1927 174 games, career pitching w. and l. 72–70, career ba .174
Marjorie Hanna	Calgary, Alberta	1944
Kay Heim (McDaniel)	Edmonton, Alberta	1943–44
Dorothy Hunter	Winnipeg, Manitoba	1943 82 games, career ba .224; chaperone for Milwaukee/Grand Rapids Chicks, 1944–54
Christine Jewitt (Beckett)	Regina, Saskatchewan	1948–49
Arleene Johnson (Noga) 'Johnnie'	Regina, Saskatchewan	1945–48 354 games, career ba .164
Marguerite Jones (Davis)	Regina, Saskatchewan	1944
Daisy (Knezovich) Junor	Regina, Saskatchewan	1946–49
Ruby Knezovich	Regina, Saskatchewan	1943–44
Mary Kustra (Shastal)	Winnipeg, Manitoba	1944
Olive (Bend) Little 'Ollie'	Poplar Point, Manitoba	1943, 1945–46 b. 1917 113 games, career pitching w. and l. 57–43, 1943 21–15, 1945 22–11, first league no-hitter, career ba .146
Lucella Maclean (Ross)	Lloydminster, Alberta	1943–44
Hazel Measner	Holdfast, Saskatchewan	1946–47
Ruth Middleton	Winnipeg, Manitoba	1950–54
Helen Nelson	Toronto, Ontario	1943
Helen Nicol (Fox) 'Nicki'	Ardley, Alberta	1943–52 b. 1920 354 games, career pitching w. and l. 163–118, 1943 31–8, 1945 24–19, career era 1.89, career ba .145
Penny O'Brien	Edmonton, Alberta	1944–45

Vickie Panos	Edmonton, Alberta	1944
Betty Petryna	Regina, Saskatchewan	1948–49
Martha Rommelaire	Edmonton, Alberta	1950
Joan Schatz	Winnipeg, Manitoba	1950–52
June Schofield	Toronto, Ontario	1948–49
Doris Shero (Witiuk)	Winnipeg, Manitoba	1950
Colleen Smith	Vancouver, British Columbia	1949
Ann Surkowski	Moose Jaw, Saskatchewan	1945
Lena Surkowski	Moose Jaw, Saskatchewan	1944–48
Anne Jane Thompson	Edmonton, Alberta	1943–44
Thelma Walmsley	Sudbury, Ontario	1946
Mildred Warwick (McAuley)	Regina, Saskatchewan	1943–44
Evelyn Wawryshyn (Moroz) 'Evie'	Tyndall, Manitoba	1946–51 b. 1924 544 games, career ba .266, on 1st all-star team in 1950; played hockey in off season
Elsie Wingrove	North Portal, Saskatchewan	1946–47
Agnes Zur(k)owski (Holmes)	Regina, Saskatchewan	1945

53 players:
Saskatchewan 24
Manitoba 11
Alberta 9
Ontario 6
British Columbia 3

Canadian Teams in Organized Baseball

Just as Canadian participation in the major leagues is a clue to the game's popularity in the country, so is a roster of Canadian teams represented in the full structure of organized baseball, the term defining the integrated corporate structure of major and minor league baseball which has been in place since the formation of the National Association of Professional Baseball Leagues in 1902. Prior to that major league baseball had other structures through which it effectively controlled the business of professional baseball. Canadian teams wanting into the professional minor leagues had to adopt a sub-servient relationship to an American-based organization. It was a compromise they were willing to make because as independent businesses it allowed them greater prosperity and an opportunity to play a role in the management of this baseball structure.

Affiliated teams were only the tip of the iceberg of Canadian baseball team development but they stand out, not only because they were mentioned in American media reports, and attracted a higher calibre of player, but as in the case of minor league teams in Toronto, Montreal, and Vancouver survived for many years as highly successful commercial operations. The first Canadian Baseball League was the 1876 organization formed in Toronto to replace the somewhat haphazard challenge and tournament play out of which previous champions had emerged. It was a completely independent Canadian entity that owed at least part of its model to the newly formed National League in the United States. Five Ontario teams from London, Guelph, Hamilton, Kingston, and Toronto were inaugural members. Within the province the league was replaced by challenge play the next year. London and Guelph however played in the American-based International Association, an organization with major league pretensions. Not only did Londoner Harry Gorman played a central role in the Association but the London Tecumsehs won the Association's first pennant.

The Tecumsehs rejoined the Association in 1878 in what would prove to be the last season of Canadian representation in an organized league comparable to those found in the United States until 1884 when a short-lived Canadian League was replaced by the Western Ontario Baseball League in mid-season. Nine teams, with some places having several representatives, played in these leagues which are the first to warrant minor league status.

Unlike Canadian ballplayers whose arrival can be traced to conditions in the game a generation before when they were growing up, Canadian teams as creatures of commerce are rooted in the everyday present. The decline of teams with a semblance of minor league identity from five in 1885 to two in 1886 and 1887 represented the impact of Canadian teams' flight from locally based associations to American-based ones which were more selective in their choice of Canadian teams.

In 1890 there were nine teams representing eight cities which could be defined as minor league. For the first time Montreal and Maritime centres in New Brunswick were included. The collapse of the

Players League after its 1890 season and the termination of the American Association shortly thereafter chilled the atmosphere for investing in baseball clubs. In 1891 Winnipeg was the only Canadian minor league representative and also the first from western Canada. Between 1892 and 1894, following the collapse of first the Players League and then the American Association, there were no Canadian minor league clubs. Toronto returned to the minors in 1895, a position it occupied continuously through the 1967 season. Teams from British Columbia made their appearance in the late 1890s though they only lasted a season. The 19th-century peak of nine achieved in 1884 and 1890 was again reached in 1899. In 1900 when not one Canadian played major league ball there were six Canadian minor league clubs, all from Quebec and Ontario. Minor league baseball flourished throughout Canada under the benign dictatorship of the National Association of Professional Baseball Leagues. By 1905 Canadian teams represented Vancouver, Victoria, Winnipeg, Montreal, as well as eight Ontario centres. For the first time in 1907 there were teams from Alberta, as four places joined the new Western Canada League. Vancouver, Winnipeg, Montreal and Ontario centres were also represented. There were 11 minor league teams in 1909, 12 a year later, 17 in 1911, 18 a year later and 24 in 1913, a number that has never been surpassed.

For the first time that season there were teams from every Canadian region. They included three from Alberta, three from Saskatchewan, two from British Columbia, one from Winnipeg, four New Brunswick clubs, Montreal, and 10 from Ontario. Twenty-one Canadian clubs made up organized baseball in the last summer before the First World War, confirming Alan Metcalfe's later conclusion about baseball's ascendant position in Canada's summer sporting calendar with the disappearance of cricket and the decline in interest in lacrosse.

Baseball suffered during the war as many parks were turned over to the military for their use. By 1918 only three teams remained. The organized game briefly flourished after the war, peaking in 1922 when 14 teams played in Calgary and Edmonton, Vancouver, four Quebec centres, and seven Ontario towns and cities. By 1926 only Toronto remained in organized baseball contrasting ironically with the almost complete disappearance of Canadians from major league line-ups. Numbers

rebounded somewhat in 1930 with eight teams but slumped again during the early Depression years to only two—in Montreal and Toronto. Minor league baseball was not significantly revived until 1937 when 13 teams existed including Nova Scotia's representatives in the Cape Breton Colliery League which lasted through 1939. The Second World War dealt another blow to Canadian baseball entrepreneurs as it would do in the United States. Only Montreal and Toronto operated between 1943 and 1945.

Throughout North America the minor leagues flourished in the years immmediately following the war, peaking in 1949 when 438 cities were members of 59 leagues. Canadian teams had remained constant at nine between 1946 and 1949 but boomed to 16 in 1951 by which time the decline was already underway throughout the United States, where television had been available a few years before its arrival in Canada. There were still 15 Canadian teams in 1954 but the pattern already witnessed in American cities was felt in Canada a year later when only eight teams survived. By the end of the decade declining attendance resulted in the departure of Montreal's International League team, one of the most successful minor league franchises of the previous two decades.

Nothing so bespeaks the decline of baseball in Canada in the 1960s as the erosion in number of Canadian teams to the one in Vancouver in 1968. Toronto's team, continuous in the minors since 1895, drew only 67,216 fans in 1967, down from a high of 446,040 in 1952. Ironically this was a period of tremendous growth in the Canadian Football League's popularity in the country, a development corresponding to the popularity of football in the United States. The most significant development in Canadian baseball history and the one which revived the game in Canada was the entry of the Montreal Expos into the National League in 1969. There was a small incease in the number of Canadian teams in organized baseball in the 1970s. The 1977 total of seven included the second Canadian major league team, the Toronto Blue Jays. Through the early 1980s Canada's two major league teams were joined by minor league franchises in Alberta and Vancouver. The 1993 total of eleven includes basically these teams together with some Ontario-based minor league squads established in part because of the success of the Toronto Blue Jays.

Canadian Team Inventory

Information Format

Province

City/Town teams listed by nickname, league and relevant years (teams in brackets are a cross section of Canadian senior teams not part of organized baseball)

Alberta

Bassano Boosters of the Western Canada League in 1912

Calgary Bronchos of the Western Canada League in 1907/ the Cowpunchers or Eyeopeners of the Western Canada League in 1909 and renamed the Bronchos year later and continue in Western Canada League until 1914/ Bronchos of Western Canada League 1920–21 and members of Western International League in 1922/ (Purity 99s, and Buffalos of Big Four League c. 1947–52)/ Stampeders of Western International League of 1953–54/ Cardinals of 'summer' Pioneer League 1977–78/ renamed Expos of 'rookie' Pioneer League 1979–84/ Cannons of Pacific Coast League 1985 through 1994

  1910 - win pennant in both halves of Western Canada League

  1912 - win 2nd half of season pennant of Western Canada League and league playoff championship

  1920 - win 2nd half of season pennant of Western Canada League and league playoff championship

  1921 - win 1st half of season pennant of Western Canada League and league playoff championship

  1922 - 1st in Western International League when it disbands 18 June

  1981 - win North Division pennant of Pioneer League

  1983 - win North Division pennant of Pioneer League

  1985 - win 1st half of North Division pennant of Pacific Coast League

  1987 - win 2nd half of North Division pennant of Pacific Coast League

  1989 - win 2nd half of North Division pennant of Pacific Coast League

  1991 - win 2nd half of North Division pennant of Pacific Coast League

Edmonton Grays of the Western Canada League in 1907/ the Eskimos of the Western Canada League 1909–11, renamed the Gray Birds a year later and continue in Western Canada League through 1913 and then renamed Eskimos and continue in Western Canada League in 1914/ Edmonton Esquimos of Western Canada League in 1920, renamed Eskimos in 1921 in Western Canada League and join Western International League in 1922/ (Eskimos and Motor Cubs of Big Four League c. 1947–52)/ Eskimos of Western International League 1953–54/ (United States college league c. late 1950s)/ Trappers of Pacific Coast League 1981 through 1994

  1983 - win 1st half of North Division pennant of Pacific Coast League

  1984 - win 1st half of North Division pennant of Pacific Coast League and league playoff championship

  1990 - win 2nd half of North Division pennant of Pacific Coast League

Lethbridge of Western Canada League in 1907/ Miners of Western Canada League 1909–11/ Expos of 'rookie' Pioneer League in 1975–76, renamed Dodgers a year later as member of 'summer' Pioneer League through 1978/ Dodgers in rookie Pioneer League 1979–83/ Mounties of 'rookie' Pioneer League 1992 through 1994

  1977 - win Pioneer League pennant

  1979 - win North Division pennant of Pioneer League and league playoff championship

  1980 - win North Division pennant of Pioneer League and League playoff championship

Medicine Hat Hatters of Western Canada League 1907/ Mad Hatters of Western Canada League 1909–mid 1910/ Hatters of Western Canada League 1913–14/ A's of 'summer' Pioneer League 1977 and renamed Blue Jays a year later continuing in rookie Pioneer League through 1994 season

> 1907 - win Western Canada League pennant
>
> 1909 - win Western Canada League pennant
>
> 1982 - win North Division pennant of Pioneer League and league playoff championship

Red Deer Eskimos of Western Canada League 1912

> 1912 - win 1st half of season pennant of Western Canada League

British Columbia

New Westminster Frasers of Northwest League 1974

Rossland of Washington State League 1897

Vancouver Horse Doctors of Northwestern League 1905 and 1907, renamed Beavers for 1908 season, continued through 1911, renamed Champions for 1912 season, renamed Bees for 1913, renamed Beavers for 1914–17/ Beavers representing Vancouver British Columbia and Vancouver Washington of Pacific Coast International League 1918/ Beavers of International Northwest League 1919, of Pacific Coast International League 1920–21, of Western International League 1922/ Vancouver Maple Leafs of Western International League 1937–38, renamed Capilanos for 1939 season through 1942/ Capilanos of Western International League 1946–54/ Mounties of Pacific Coast League 1956–62/ Mounties of Pacific Coast League 1965–69/ Canadians of Pacific Coast League 1978 through 1994

> 1908 - win Northwestern League pennant
>
> 1911 - win Northwestern League pennant
>
> 1913 - win Northwestern League pennant
>
> 1914 - win Northwestern League pennant
>
> 1919 - 1st in International Northwest League when it disbands on 7 June
>
> 1942 - win Western International League pennant
>
> 1947 - win Western International League pennant

> 1949 - win Western International League playoff championship
>
> 1954 - win 1st half of season pennant of Western Interantional League and league playoff championships
>
> 1979 - win 2nd half of North Division pennant of Pacific Coast League
>
> 1980 - win 2nd half of North Division pennant of Pacific Coast League
>
> 1985 - win 2nd half of North Division pennant of Pacific Coast League and league playoff championship
>
> 1986 - win both halves of North Division pennant of Pacific Coast League
>
> 1988 - win both halves of North Division pennant of Pacific Coast League
>
> 1989 - win 1st half of North Division pennant of Pacific Coast League and league playoff championship
>
> 1992 - win 1st half of North Division pennant of Pacific Coast League
>
> 1994 - win both halves of North Division pennant of Pacific Coast League

Victoria (Amity 19th century)/ Chappies of Pacific Northwest League 1896/ Legislators of Northwestern League 1905/ Bees of Northwestern League 1911–15/ Islanders of International Northwest League 1919, of Pacific Coast International League 1920, renamed Bees of Pacific Coast International League 1921/ Athletics of Western International League 1946–51, renamed Tyees for 1952 season continued through 1954/ Mussels of Northwest League 1978–80

> 1920 - win Pacific Coast International League pennant
>
> 1952 - win Western International League pennant

Manitoba

Brandon (of the Manitoba Senior Baseball League 1903)/ Angels of the Northern League 1908/

Western Canada League 1909–11/ Greys of Northern League in 1933/ (Greys of Mandak League circa post-World War Two to early fifties)

1908 - win Northern League pennant

1933 - win 2nd half of season pennant of Northern League

St Boniface of Northern League 1915

Winnipeg (CPR, Metropolitans, and Hotel teams of Manitoba Baseball League 1886/ of Red River Valley League in 1891/ Maroons aka Peggers of Northern League 1902–05/ of North Copper Country League 1906–07/ of Northern League 1908/ of Western Canada League 1909–11/ of Central International League 1912/ of Northern League 1913–17/ Maroons of Western Canada League 1919–21/ Maroons of Northern League 1933–42/ (Buffalos, and Elmwood Giants of Mandak League circa post-World War Two to early fifties)/ Gold Eyes of Northern League 1954–64/ Gold Eyes of Northern League 1969/ Whips of International League 1970–71/(Goldeyes of Northern League 1994)

1903 - win Northern League pennant

1907 - win North Copper Country League pennant

1916 - win Northern League pennant

1919 - win 1st half of season pennant of Western Canada League

1921 - win 2nd half of season pennant of Western Canada League

1935 - win 1st of season pennant of Northern League and league playoff championships

1939 - win Northern League pennant and league playoff championships

1942 - win Northern League playoff championship by forfeit

1957 - win 2nd half of season pennant of Northern League and league playoff championship

1959 - win Northern League pennant and league playoff championship

1960 - win Northern League pennant and league playoff championship

1994 - win independent Northern League pennant and league playoff championship

New Brunswick

St Stephen Downeasters in alliance with Calais, Maine in the New Brunswick-Maine League 1913/(of the St Croix Baseball League c. 1920s)/(Mohawks of the York-Charlotte League 1930 and Mohawks of the York-Charlotte-Carleton Senior Baseball League 1931, renamed Kiwanis and play as independent team 1932–35, renamed St Croixs and play as independent team 1936–39)

Fredericton of New Brunswick League 1890/Pets of New Brunswick-Maine League 1913

1913 - win New Brunswick-Maine League Pennant

Moncton of New Brunswick League 1890

St Croix of New Brunswick-Maine League 1913

Saint John of New Brunswick League 1890/Shamrocks of New Brunswick League 1890/Marathons of the New Brunswick-Maine League 1913

1890 - Shamrocks win New Brunswick League pennant

Newfoundland

Bell Island (of Inter-Town League 1913)

Grand Falls (of Inter-Town League 1913)

St John's (Red Lions, Wanderers, Benevolent Irish Society, and Shamrocks of St John's Amateur Baseball League 1913)/(St John's of Inter-Town League 1913)

1913 - Wanderers win St John's Amateur League pennant

1913 - St John's win Inter-Town League pennant

Nova Scotia

Dartmouth (Arrows c. 1940s)

Dominion Hawks of Cape Breton Colliery League 1937–38

Glace Bay Miners of Cape Breton Colliery League 1937–39

> 1937 - win Cape Breton Colliery League pennant

> 1938 - win Cape Breton Colliery League pennant, and league playoff championships over New Waterford

Halifax (Shipyards *c.* 1940s/Cardinals *c.* 1950s)

Kentville (Wildcats, *c.* 1950s)

Liverpool (Larrupers *c.* 1930s)

New Waterford Dodgers of Cape Breton Colliery League 1937–39

Springhill (Fencebusters *c.* 1930s)

Stellarton (Albions *c.* 1950s)

Sydney Mines Ramblers of Cape Breton Colliery League 1937–39

Sydney Steel Citians of Cape Breton Colliery League 1937–39

> 1937 - win Cape Breton Colliery League playoff championships over Glace Bay

> 1939 - win Cape Breton Colliery League pennant, and league playoff championships over New Waterford

Truro (Bearcats *c.* 1940s)

Yarmouth (Gateways *c.* 1930s)

Ontario

Belleville of the Eastern International 1888

Berlin/Kitchener [name of city changed from Berlin to Kitchener during First World War] Berlin Green Sox of the Canadian League 1911, renamed Berlin Busy Bees 1912–13 of Canadian League/Kitchener Beavers of Michigan-Ontario League 1919–21, renamed Kitchener Terriers of Michigan-Ontario League 1922/Kitchener Colts of Michigan-Ontario League 1925

> 1911 - win Canadian League pennant

Bowmanville (Royal Oaks *c.* 1870s)

Brantford Indians of Canadian League 1905/Red Sox of Canadian League 1911–1915/Red Sox of Michigan-Ontario League 1919–21, renamed Brants of Michigan-Ontario League 1922/Red Sox of Ontario League 1930

Brockville of Eastern International League mid 1888/Pirates of Canadian-American League mid 1936, renamed Blues of Canadian-American League 1937

Chatham Babes of Canadian League mid 1898, renamed Reds of Canadian League 1899 and International League 1900

Cornwall Bisons of Canadian-American League 1938, renamed Maple Leafs of Canadian-American League 1939/Canadians of Border League 1951

> 1938 - leading Canadian-American League playoff championships 2–1 (with one tie) when bad weather ends season

Fort William and Port Arthur/Thunder Bay [Fort William and Port Arthur amalgamated as Thunder Bay in 1970] Fort William Canadians of Northern League 1914–15, renamed Fort William-Port Arthur Canadians of Northern League 1916/(Thunder Bay Whiskey Jacks of Northern League 1993 through 1994)

Galt of Canadian League 1896/of Canadian League 1914

Guelph Maple Leafs of Canadian League 1876/Maple Leafs of International Association 1877/Maple Leafs of Ontario League 1884/of Canadian League 1885/of Canadian League 1896–97/of Canadian League 1899/Biltmores of Canadian League 1905/Maple Leafs of Canadian League 1911–13, and 1915/Biltmores of Ontario League 1930

> 1869 through 1875 - Maple Leafs win Ontario challenge championship (Silver Ball competition)

1894 - Ontario champions

1895 - Ontario champions

1896 - win Canadian League pennant

Hamilton Standards of Canadian League 1876/ Clippers of Ontario League (also referred to as Canadian League) 1884/Primroses of Ontario League (also referred to as Canadian League) 1884/Clippers of Canadian League 1885/Primroses of Canadian League 1885/Clippers of Canadian League 1885/ Clippers of International League 1886, renamed Hams of International League 1887 and International Association 1888–89, renamed Blackbirds in International League 1890/Hams, aka Blackbirds, of Canadian League 1896–99 and International League 1900/Hams of International League 1908/Kolts of Canadian League 1911–14, renamed Hams mid 1914–15/Tigers of International League 1918/Tigers of Michigan-Ontario League 1919–23, renamed Clippers 1924–25/Tigers of Ontario League 1930/Red Wings of PONY League 1939–42/Cardinals of PONY League 1946–56/ Redbirds of New York-Pennsylvania (New York-Penn) League 1988–92

1884 - Clippers win Ontario League championship

1885 - Clippers win Canadian League pennant

1897 - win Canadian League pennant

1908 - leading International League when league disbands in late July

1922 - win 2nd half of season pennant of Michigan-Ontario League, and league playoff championship

1925 - win first half of season pennant of Michigan-Ontario League

1952 - win PONY League pennant

1955 - win PONY League pennant and league playoff championship

1992 - win Stedler Division pennant of New York-Penn League

Ingersoll of Canadian League 1905

1868 - Victorias win Ontario challenge championship (Silver Ball competition)

Kingston St Lawrence of Canadian League 1876/of

Eastern International League 1888/Ponies of Border League 1946–51

1951 - leading Border League when league disbands 16 July

London Tecumsehs of Canadian League 1876/ Tecumsehs of International Association 1877–78/ Alerts of Ontario League 1884/Londons of Ontario League 1884/Atlantics of Ontario League 1884/ Londons of Canadian League 1885/Tecumsehs of International Association 1888–89 and International League 1890/Alerts of Canadian League 1896–97, renamed Tecumsehs 1898–1900/of International League 1900/Cockneys of International League 1908/Cockneys of Canadian League 1911 renamed Tecumsehs 1912–15/Tecumsehs of Michigan-Ontario League 1919–24, renamed Indians 1925/Tecumsehs of Ontario League 1930/Pirates of PONY League 1940–41/Tigers of Eastern League 1989–93

1876 - win Canadian League pennant

1877 - win International Association pennant

1899 - win pennant in both halves of Canadian League season

1900 - win shortened-season pennant of International League

1920 - win Michigan-Ontario League pennant

1921 - win 2nd half of season pennant of Michigan-Ontario League, and league playoff championship, and Class B championship

1925 - win 2nd half of season pennant of Michigan-Ontario League, and league playoff championship

1930 - leading 2nd half of season pennant of Ontario League when league disbands 22 July

1990 - win Eastern League playoff championship

Newcastle (Beavers *c.* 1870s)

Ottawa of Eastern International League 1895/of Eastern League mid 1898/of Canadian League 1905/Senators of Canadian League 1912–15/

Canadiens of Eastern Canada League 1922–23/ Ottawa-Hull Senators of Quebec-Ontario-Vermont League 1924/Senators of Canadian-American League 1936, renamed Braves 1937–38, renamed Senators 1939/Ottawa-Ogdensburg (New York) Senators of Canadian-American League 1940/Nationals of Border League 1947, renamed Senators 1948–49/Nationals of Border League 1950/Giants of International League 1951, renamed Athletics 1952–54/Lynx of International League 1993 through 1994

   1912 - win Canadian League pennant

   1913 - win Canadian League pennant

   1914 - win Canadian League pennant

   1915 - win Canadian League pennant

   1940 - win Canadian-American league pennant

   1947 - win Border League pennant and league playoff championship

   1948 - win Border League pennant

   1950 - win Border League pennant

Perth Blue Cats, aka Royals and Braves, of Canadian-American League 1936/Perth-Cornwall Grays of Canadian-American League 1937

   1936 - win Canadian-American League pennant, and league playoff championships over Brockville

   1937 - win Canadian-American League pennant

Peterborough Whitecaps of Canadian League 1912, renamed Petes 1913–14

Petrolia Imperials of Ontario League 1884

St Catharines Brewers of Ontario League 1930/Blue Jays of New York-Pennsylvania (New York-Penn) League 1986 through 1994

   1986 - win Wrigley Division pennant of New York-Penn League and league playoff championship

   1993 - win Stedler Division pennant of New York-Penn League

St Thomas Saints of Canadian League 1898–99/of Canadian League 1905/of International League

1908/Saints of Canadian League 1911–15/Blue Sox of Ontario League 1930

   1898 - win Canadian League pennant

Sarnia, Port Huron, Michigan/Sarnia, Ontario Saints of Michigan-Ontario League 1922

Sault Ste Marie Soos of Copper County League 1905

Smiths Falls Beavers of Canadian-American League 1937

Stratford Poets of Canadian League 1899

Toronto of Canadian League 1876/of Ontario League 1884/of Canadian League 1885/Canucks of International League 1886–87, of International Association 1888–89, of International League 1890/ Canadians, aka Red Stockings, of Eastern League 1895, renamed Canucks 1896–1900 of Eastern League, renamed Royals 1901, renamed Maple Leafs of Eastern League 1902–11, of International League 1912–67 [league called New International League 1918–19]/Beavers of Canadian League 1914/(Maple Leafs of Inter-County Major Baseball League 1969 through 1994)/Blue Jays of American League 1977 through 1994

   1887 - win International League pennant

   1902 - win Eastern League pennant

   1907 - win Eastern League pennant and Junior World Series

   1912 - win International League pennant

   1917 - win International League pennant, but lose Junior World Series

   1918 - win New International League pennant

   1926 - win International League pennant and Junior World Series

   1934 - win International League playoff championship, but lose Junior World Series

   1943 - win International League pennant

   1954 - win International League pennant

   1956 - win International League pennant

   1957 - win International League pennant

   1960 - win International League pennant and

league playoff championship, but lose Junior World Series

1965 - win International League playoff championship

1966 - win International League playoff championship

1985 - win East Division pennant of American League, but lose American League playoff championship

1989 - win East Division pennant of American League, but lose American League playoff championship

1991 - win East Division pennant of American League, but lose American League playoff championship

1992 - win East Division pennant of American League and American League playoff championship and World Series

1993 - win East Division pennant of American League and American League playoff championship and World Series

Welland Pirates of New York-Pennsylvania (New York-Penn) League 1989 through 1994

Windsor of Border League 1912–13/of Eastern Michigan League 1914

Woodstock of Ontario League 1884/Bains of Canadian League mid 1899/ Maroons of Canadian League 1905

1865 through 1868 - Young Canadians win Ontario challenge championship (Silver Ball competition)

1905 - winners of Canadian League pennant

Prince Edward Island

Charlottetown (Abegweits *c.* 1920s)

Summerside (Curran and Briggs *c.* 1940s)

Quebec

Cap de la Madeleine Madcaps of Eastern Canada League 1922

Drummondville Tigers of Quebec Provincial League 1940/Cubs of Provincial League 1950–52, renamed Royals of Provincial League 1953, renamed A's of Provincial League 1954

Farnham Pirates of Provincial League 1950–51

Granby Red Socks of Quebec Provincial League 1940/Red Sox of Border League 1946/Red Sox of Provincial League 1950–51, renamed Phillies of Provincial League 1952–53

Montreal of International League mid 1890 [formerly Hamilton Redbirds]/of International League 1890 [formerly in Buffalo, went to Montreal, and then Grand Rapids, Michigan]/of Eastern International League 1895/Jingos of Eastern League 1897–1900, renamed Royals of Eastern League 1901–11, of International League 1912–17/Royals of Eastern Canada League 1922–23, of Quebec-Ontario-Vermont League 1924/Canadiens of Eastern Canada League 1923, of Quebec-Ontario-Vermont League 1924/Royals of International League 1928–60/Expos of National League 1969 through 1994

1898 - win Eastern League pennant

1923 - Royals win 2nd half of season pennant of Eastern Canada League and league playoff championship

- Canadiens win 1st half of season pennant of Eastern Canada League

1935 - win International League pennant

1941 - win International League playoff championship and lose Junior World Series

1945 - win International League pennant

1946 - win International League pennant, league playoff championship, and Junior World Series

1948 - win International League pennant, and league playoff championship, and Junior World Series

1949 - win International League playoffs, but lose Junior World Series

1951 - win International League pennant, and league playoff championship, but lose Junior World Series

1952 - win International League pennant

1953 - win International League playoff championship and Junior World Series

1955 - win International League pennant

1958 - win International League pennant and league playoff championship, but lose Junior World Series

1981 - win 2nd half of season pennant of East Division of National League, win East Division playoff championship, and lose National League playoff championship

1994 - win East Division pennant of National League with best record in major league baseball at time of season ending strike

Outremont Canadiens of Quebec-Ontario-Vermont League 1924

Quebec City of International League 1890 (according to Filichia but not confirmed in International League records)/Bulldogs of Eastern Canada League 1923, of Quebec-Ontario-Vermont League 1924/Athletics of Quebec Provincial League 1940, of Canadian-American League 1941–42/Alouettes, aka Larks, of Canadian-American League 1946–48, renamed Braves of Canadian-American League 1949–50, of Provincial League 1951–55/Carnavals of Eastern League 1971–mid 75, renamed Metros mid 1975–77

1924 - win both halves of season pennants of Quebec-Ontario-Vermont League

1949 - win Canadian-American League pennant and league playoff championship

1950 - win Canadian-American League pennant and league playoff championship

1952 - win Provincial League playoff championship

1953 - win Provincial League playoff championship

1954 - win Provincial League pennant and league playoff championship

1955 - win Provincial League playoff championship

1974 - win National Division pennant of Eastern League

St-Hyacinthe Saints of Provincial League 1940/Saints of Provincial League 1950–51, renamed A's 1952–53

1940 - win Quebec Provincial League pennant

1952 - win Provincial League pennant

St Jean Braves of Provincial League 1950–51, renamed Canadians 1952–55

1950 - win Provincial League pennant and league playoff championship

1955 - win Provincial League pennant

Sherbrooke Braves of Quebec Provincial League 1940/Canadiens of Border League 1946/Athletics of Provincial League 1950–51/Indians of Provincial League 1953–55/Pirates of Eastern League 1972–73

1951 - win Provincial League pennant and league playoff championship

1953 - win Provincial League pennant

Thetford Mines Miners of Provincial League 1953–55/Pirates of Eastern League 1974, renamed Miners 1975

1974 - win Eastern League playoff championship

Trois Rivières [or Three Rivers] Trios of Eastern Canada League 1922–23/Foxes of Quebec Provincial League 1940, renamed Renards of Canadian-American League 1941, renamed Foxes 1942/Royals of Canadian-American League 1946–50, of Provincial League 1951, renamed Yankees 1952–53, renamed Phillies 1954/Aigles of Eastern League 1971–77

1922 - win Eastern Canada League pennant

1940 - win Quebec Provincial League playoff championships over Granby

1946 - win Canadian-American League pennant and league playoff championship

1971 - win National Division pennant of Eastern League

1972 - win North Division pennant of Eastern League

1976 - win North Division pennant of Eastern League

1977 - win CAN-AM Division pennant of Eastern League

Valleyfield of Eastern Canada League 1922

Saskatchewan

Moose Jaw Robin Hoods of Western Canada League 1909–11/Robin Hoods of Western Canada League 1913–14/Robin Hoods of Western Canada League 1919–20, renamed Millers 1921

1911 - win Western Canada League pennant

1913 - win 2nd half of season pennant of Western Canada League and league play-off championship

Regina Bonepilers of Western Canada League 1909–10/Red Sox of Western Canada League 1913–14/

Senators of Western Canada League 1919–21/ (Cyclones of independent North Central League 1994)

1920 - win 1st half of season pennant of Western Canada League

1994 - win West Division pennant of independent North Central League

Saskatoon Berry Pickers of Western Canada League 1910, renamed Quakers 1911/Quakers of Western Canada League 1913, renamed Sheiks 1914/Quakers of Western Canada League 1919–21/(Riot of independent North Central League 1994)

1913 - win 1st half of season pennant of Western Canada League

1914 - win Western Canada League pennant

1919 - wins 2nd half of season pennant of Western Canada League and league play-off championship

APPENDIX D

Asahi Roster (1914–1941)

The following list of Japanese Canadian members of the Asahi baseball team of Vancouver was compiled at the time of their reunion in 1972. It demonstrates the extent to which they dispersed throughout the country following the internment years of World War II, and why the team never reformed afterwards.

PLAYER (Position)	Playing Period	1972 Hometown
Endo, K. (p)	1938	Winnipeg
Fukuda (of)	1930-31	Japan
Fukui, Jim (p)	1937-38	Vancouver
Fukui, Joe (1b)	1928-38	Vancouver
Furumoto, Ted (p)	1914-25	Japan
Hayami, Jack (p)	1921	Montreal
Horii, Yo (c)	1914-26	Deceased
Inouye, Tokuichi (inf)	1919	Deceased
Ito, George (3b)	1918-26	Montreal
Kaminishi, Koichi (inf)	1939-40	Kamloops
Kasahara, B.	1918-23	Deceased
Kato, Fred (p)	1921	Deceased
Kato, George (p)	1925-29	Deceased
Kitagawa, Eddie (of)	1917-31	Toronto
Kitagawa, Mickey (p)	1914-25	Deceased
Kodama (of)	1914-17	Deceased
Korenaga, Abe S. (of)	1933-38	Deceased
Korenaga, R. (p)	1938	Toronto
Kutsukake, Ken (c)	1938-41	Toronto
Kutsukake, Ray (p)	1933	Toronto
Kitamura, Ken (inf)	1918	Toronto
Maikawa, Mickey (p)	1928-34	Toronto
Maruno, Mike (ss)	1935-41	Toronto
Masuda, Mousie (c, of)	1933-38	Deceased

Matoba, Tom	1914-30	Toronto
Matsumiya, Sota (of)	1914-23	Japan
Miike, Muneo (of)	1938	Toronto
Mitsui, Koei (of, c)	1939-41	Deceased
Miyata, Tom (of)	1922-28	Japan
Miyasaki, Harry (1b)	1918-22	Deceased
Miyasaki, Yoshio	1925	Japan
Nakamura, Ed. (of)	1934-41	Toronto
Nakamura, Frank (2b)	1923-34	Toronto
Nakamura, Sally (2b)	1926-32	Japan
Nakanishi, Ken (p)	1933	Deceased
Niimi, Joe (of)	1918-21	Japan
Niimi, Toragoro	1919	Richmond, B.C.
Nishidera, Roy (p)	1926-30	Japan
Nishihara, Nag. (p)	1932-41	Princeton, B.C.
Noda, Ken (p)	1933-35	Deceased
Oda, B. (c)	1921-22	Toronto
Omoto, Tashiro	1929	Kapuskasing
Sawayama, Tom (p)	1939-41	Toronto
Sato, Mickey (of, p)	1925-31	Deceased
Sekine, Jubo (of)	1928-29	Toronto
Shima	1914-	Deceased
Shimada, Ken (inf)	1932-33	Toronto
Shiraishi, Frank (of)	1930-41	Hamilton
Shishido, George (2b)	1932-41	Toronto
Suga, Kaz (of)	1933-41	Montreal
Suga, Ty (p)	1925-39	Montreal
Suzuki, Ken (of)	1914-16	Deceased
Tabata	1914	Deceased
Tanaka, Charlie (1b)	1922-27	Toronto
Tanaka, Geo. (p)	1922-33	Toronto
Tanaka, Herbie (3b)	1928-37	Montreal
Tanaka, T.H. (p)	1921	Toronto
Terada, Chuck (3b)	1940	Seattle
Terakita, Mickey (p)	1929-30	Vancouver
Uchiyama, Yuji	1918-21	Deceased
Uno, Yuki (1b)	1937-41	Montreal
Yamamura, Ken (inf)	1923-31	Toronto
Yamamura, Roy (ss)	1923-41	Toronto
Yasui, Bob H. (of)	1928-33	Toronto
Yasui, Reggie H. (c)	1923-37	Toronto
Yonemoto, Ross (inf)	1921-22	Toronto
Yoshinaka, George (p)	1939-40	Lethbridge
Yoshioka, Happy (p)	1918	Japan
Yoshioka, Sutejiro (inf)	1923-24	Kamloops

Asahi Managers

Miyasaki, Matsujiro	1914-1917
Kasahara, B.	1918-1921
Miyasaki, Harry	1922-1929
Kitagawa, Eddie	1930-1931
Tanaka, George	1932-1933
Nakamura, Frank	1934
Yasui, Reggie H.	1935-1937
Yamamura, Roy	1938-1941

APPENDIX E

Expos and Blue Jays Record

Montreal Expos Historical Record 1969-94

14th best all-time winning percentage of existing major league franchises

Year	Season Record		Cumulative Record	
1969	52-110	.321	52-110	= .321
1970	73-89	.451	125-199	= .385
1971	71-90	.441	196-289	= .404
1972	70-86	.449	266-375	= .414
1973	79-83	.488	345-458	= .429
1974	79-82	.491	424-540	= .439
1975	75-87	.463	499-627	= .443
1976	55-107	.340	554-734	= .430
1977	75-87	.463	629-821	= .433
1978	76-86	.469	705-907	= .437
1979	95-65	.594	800-972	= .451
1980	90-72	.556	890-1044	= .460
1981	60-48	.556	950-1092	= .465
1982	86-76	.531	1036-1168	= .470
1983	82-80	.506	1118-1248	= .472
1984	78-83	.484	1196-1331	= .473
1985	84-77	.522	1280-1408	= .476
1986	78-83	.484	1358-1491	= .476
1987	91-71	.562	1449-1562	= .481
1988	81-81	.500	1530-1643	= .482
1989	81-81	.500	1611-1724	= .483
1990	85-77	.525	1696-1801	= .485
1991	71-90	.441	1767-1891	= .483
1992	87-75	.537	1854-1966	= .485
1993	94-68	.580	1948-2034	= .489
1994	74-40	.649	2022-2074	= .493

Toronto Blue Jays Historical Record 1977-94

13th best all-time winning percentage of existing major league franchises

Year	Season Record		Cumulative Record
1977	54-107	.335	54-107 = .335
1978	59-102	.366	113-209 = .350
1979	53-109	.327	166-318 = .343
1980	67-95	.414	233-413 = .360
1981	37-69	.349	270-482 = .359
1982	78-84	.481	348-566 = .380
1983	89-73	.549	437-639 = .406
1984	89-73	.549	526-712 = .424
1985	99-62	.615	625-774 = .446
1986	86-76	.531	711-850 = .455
1987	96-66	.593	807-916 = .468
1988	87-75	.537	894-991 = .474
1989	89-73	.549	983-1064 = .480
1990	86-76	.531	1069-1140 = .484
1991	91-71	.562	1160-1211 = .489
1992	96-66	.593	1256-1277 = .495
1993	95-67	.586	1351-1344 = .501
1994	55-60	.478	1406-1404 = .500

BIBLIOGRAPHY

Chapter 1
Canada in the Country of Baseball

Books consulted in this section were used as reference works throughout this text. They included *The Canadian Encyclopedia* 2nd ed. (Edmonton: Hurtig, 1988), Thorn and Palmer's *Total Baseball: The Ultimate Encyclopedia of Baseball*, 3rd ed. (New York: Harper Perennial, 1993), *The Baseball Encyclopedia* (ed. by Reichler for Macmillan Publishing, several editions), Peter Filichia's *Professional Baseball Franchises* (New York: Facts on File, 1993), Johnson and Wolff's *The Encyclopedia of Minor League Baseball* (Durham, North Carolina: Baseball America, Inc., 1993), Morrow, Keyes, Simpson, Cosentino and Lappage's *A Concise History of Sport in Canada* (Toronto: Oxford University Press, 1989), Howell and Howell's *Sports and Games in Canadian Life* (Toronto: Macmillan, 1969), Alan Metcalfe's *Canada Learns to Play: The Emergence of Organized Sport, 1807-1914* (Toronto: McClelland and Stewart, 1987), Samuel Moffett's *The Americanization of Canada* (Toronto: University of Toronto Press, c. 1972, originally published in 1907).

Articles consulted included Ojala and Gadwood's 'The Geography of Major League Baseball Player Production: 1876-1989' from the *Minneapolis Review of Baseball* 10, 1 (1991), Goldwin Smith's *The Bystander* (August 1880) on 'National Games', George Bowering's 'Baseball and the Canadian Imagination' from *Imaginary Hand: Essays by George Bowering* (Edmonton: NeWest Press, 1988).

Correspondence and additional materials provided by the Canadian Federation of Amateur Baseball in Ottawa, Jack Stott of Hamilton on the 1886 baseball on rollers season in Hamilton, John Fogg on the Toronto Garrison Indoor Fastball League at Toronto's Fort York Armories and University Armories.

Chapter 2
Early Baseball in Canada

Books consulted included Melvin Adelman's *A Sporting Time: New York City and the Rise of Modern Athletics, 1820-70* (Chicago: University of Illinois Press, 1986), Harold Seymour's *Baseball*, 2 vols (New York: Oxford University Press, 1960, 1971), *Beachville: The Birth Place of Oxford 1784-1967* (Beachville Centennial Committee, 1967) *Beachville Cemetery, West Oxford Township* (Ontario Genealogical Society, Oxford County Branch, no date), *A History of Brighton: Being the Story of a Woodstock Settlement from the Early Thirties* (Oxford Museum Bulletin No. 7, no date), Fred Landon's *Western Ontario and the American Frontier* (Toronto: Ryerson Press, 1941), Wilson Green's *Red River Revelations* (Winnipeg: Hignell Printing, 1974).

Additional texts consulted are listed below in the Maritimes section under authors Flood, Payzant, Martin, and O'Neill.

Articles consulted included Bouchier and Barney's 'A Critical Examination of a Source on Early Ontario Baseball: The Reminiscence of Adam E. Ford' from the *Journal of Sport History* 15, 1 (Spring 1988), Robert Barney's 'Diamond Rituals: Baseball in

Canadian Culture' from *Baseball History 2* (Westport CT: Meckler Books, 1989), Robert Barney's 'In Search of a Canadian Cooperstown: The Future of the Canadian Baseball Hall of Fame' from *Nine: A Journal of Baseball History and Social Policy* 1, 1 (Fall 1992), Fred Landon's 'The Common Man in the Era of the Rebellion in Upper Canada' from *Aspects of Nineteenth-Century Ontario* (Toronto: University of Toronto Press, 1974), James Robert Anderson's 'Notes and Comments on Early Days and Events in British Columbia' (from the Archives of the Province of British Columbia), Adam Ford's letter to *The Sporting Life* published 5 May 1886 describing the Beachville game, Melvin Adelman's 'Premature Modernization and the Failure of Cricket in America: The New York Experience, 1840-1865' as presented to the conference of the North American Society for Sport History in 1980.

Correspondence and additional materials include a 1981 letter from Wilson Green of Winnipeg to the author describing the game of 'bat' as likely played in the Red River Settlement, a 1985 letter from Wayne McKell of St Chrysostome, Quebec to the author describing a game in the late 1830s in Huntingdon, Quebec as recorded in a collection of notes taken down by Robert Sellar, whose later 1907 self-published work *The Tragedy of Quebec: The Expulsion of its Protestant Farmers* was one of the most controversial politico-religious tracts ever circulated in Canada.

Chapter 3
19th Century Ontario Baseball

Books consulted included Zane Grey's 'The Winning Ball' from *The Red Headed Outfield and Other Baseball Stories* (New York: Grosset and Dunlap, 1920), Ted Vincent's *Mudville's Revenge: The Rise and Fall of American Sport* (New York: Seaview Books, 1981), Leo Johnson's *History of Guelph 1827-1927* (Guelph: Guelph Historical Society, 1977), *A Cyclopedia of Canadian Biography* (1886), *Historical Atlas of Wellington County* (Toronto: 1906), *Bryce's Canadian Baseball Guide* (London: William Bryce, 1876), Bryan D. Palmer's *A Culture in Conflict: Skilled Workers and Industrial Capitalism in Hamilton, Ontario 1860-1914* (Montreal: McGill-Queen's University Press, 1979), Whipp and Phelps, *Petrolia 1866-1966* (Petrolia: Petrolia Advertiser-Topic, 1966), D.L. Cosens' *The Donnelly Tragedy 1880-1980* (London: Phelps Publishing, 1980), Edwin Guillet's *Early Life in Upper Canada* (Toronto: University of Toronto Press), Charles Bruce's *News and the Southams* (Toronto: Macmillan, 1968).

Articles consulted included Lisa Bowes, 'George Sleeman and the Brewing of Baseball in Guelph 1872-1886' from *Historic Guelph* magazine of the Guelph Historical Society (October 1988), Keith Heidorn's 'Diamonds in the Rough: Baseball in Canada 1860-1890' from *Early Canadian Life* (May 1979), David Bernard's 'The Guelph Maple Leafs: A Cultural Indicator of Southern Ontario' from *Ontario History* LXXXIV, 3 (September 1992), Armstrong and Brock's 'The Rise of London: A Study of Urban Evolution in Nineteenth-Century Southwestern Ontario' from *Aspects of Nineteenth-Century Ontario* (Toronto: University of Toronto Press, 1974), Hugh Grant's 'The "Mysterious" Jacob L. Englehart and the Early Ontario Petroleum Industry' from *Ontario History* LXXXV, 1 (March 1993).

Newspapers consulted regarding 19th-century baseball in Ontario included the *Hamilton Spectator*, the *Guelph Evening Mercury*, the *Guelph Daily Mercury*, the *London Advertiser*, the *London Free Press*.

Additional written and pictorial sources included the files of the Guelph Civic Museum, the Verne McIlwraith Collection in the Guelph Public Library, the Sleeman Papers formerly of the University of Western Ontario Regional Collection and now located in the University of Guelph Archives at the McLaughlin Library, the London Public Library, *Nineteenth Century Notes*, a newsletter of the Nineteenth Century Committee of the Society for American Baseball Research.

Of significant note in the study of baseball in London, Ontario is the baseball coverage for over a century by three splendid local journalists in the pages of the *London Free Press* and the *London Advertiser*. Orlo Miller's *A Century of Western Ontario: The Story of London, 'The Free Press', and Western Ontario, 1849-1949* (Westport, Connecticut: Greenwood Press, 1972) sheds particular light on newspaperman and London Tecumsehs' general manager Harry Gorman who ensured that baseball was generously reported in the great London baseball era of the 1870s. Journalist Frank Adams began at the *London Advertiser* in 1880; his articles on baseball in London continued over a half century and included a report on the London Tecumsehs Minute Book in the *London Advertiser* (7 October 1935), 'Sundry Memories of Old-Time Baseball' in the *London Free Press* (19 May 1938), and 'Some Baseball History, Both Amateur and Professional, in the City of London' in the *Canadian Science Digest* (August 1938). Les Bronson continued the tradition of documenting the city's baseball history. He was still writing during the franchise life of the London Tigers (1989-93) and his paper on the London Tecumsehs ball club presented to the London-Middlesex Historical Society 15 February 1972 now resides in the University of Western Ontario Regional Collection.

Correspondence and additional materials provided by among others Les Bronson on early London baseball, and Joseph Overfield of Buffalo on the International Association including Harry Gorman's letter to Buffalo management prior to the 1878 season indicating the National League's interest in having London as a member.

Chapter 4
Maritime Baseball

Books consulted included A.J. 'Sandy' Young's *Beyond Heroes: A Sport History of Nova Scotia* (Hantsport, N.S.: Lancelot Press, 1988) with chapters on baseball, softball, and cricket, Robert Ashe's *Even the Babe Came to Play: Small Town Baseball in the Dirty 30s* (Halifax: Nimbus Publishing, 1991), the story of senior baseball in St Stephen, New Brunswick in the 1930s; Phyllis Blakeley's *Glimpses of Halifax* (Belleville: Mika Publishing, 1973) describes 19th-century baseball in Halifax, Brian Flood's *Saint John: A Sporting Tradition* (Saint John: Neptune Publishing, 1985) covers the entire history of baseball in the New Brunswick city, Paul O'Neill's *The Oldest City: The Story of St John's Newfoundland* (Don Mills: Musson Book Co., 1975) provides a brief survey of baseball and cricket in St John's, Burton Russell's *Looking Back—A Historical Review of Nova Scotia Senior Baseball 1946-1972* (Kentville, 1973); John Patrick Martin's *The Story of Dartmouth* (Dartmouth: 1957) describes the first bat and ball game in Dartmouth as mentioned in perhaps the first newspaper account of baseball play in Canada in Joseph Howe's *The Novascotian* on 1 July 1841; Joan Payzant's *Halifax: Cornerstone of Canada* (Halifax: Windsor Publications, 1985) briefly mentions early baseball in Halifax in 1838; H. Charles Ballem's

Abegweit Dynasty 1899-1954: The Story of the Abegweit Amateur Athletic Association (Summerside, PEI: Williams and Crue, 1986).

Articles consulted include Colin Howell's 'Baseball, Class and Community in the Maritime Provinces, 1870-1910' from *Histoire sociale/Social History* (November 1989), Colin Howell's 'Baseball and the Native Peoples of Atlantic Canada and Maine' presented at the 1993 Conference of the North American Society for Sport History, David Folster's 'The Old Home Team' on baseball in Grand Falls, New Brunswick from *The Atlantic Advocate* (July 1968), Bill Ledwell on 'Charlottetown's Vern Handrahan' from the *Charlottetown Monthly Magazine* (October-November 1984), Robert Ashe's 'The Boys of Summer, 1931' from *Atlantic Insight* (August/September 1981) on the St Stephen-Milltown team.

Newspaper sources included the *Charlottetown Examiner* for information on 19th-century baseball in Prince Edward Island, *The Halifax Evening Express* and the *Morning Herald* of Halifax for information on 19th-century baseball in Halifax; for 19th-century baseball in New Brunswick, newspapers consulted included the *Morning Journal* of Saint John, the *Morning Freeman*, the *Saint John Daily News*, the *New Brunswick Reporter* of Fredericton; the *Daily News* of St John's for information on baseball in Newfoundland before World War One. The *New York Clipper* was consulted for information on Maritime-born ballplayers of the 19th century. Information on Rob Butler's Newfoundland roots appeared in Christie Blatchford's *Toronto Sunday Sun* column (10 October 1993). The *Cape Breton Post* (22 September 1979) recalled an umpire's experience in the Cape Breton Colliery League.

Additional written sources included the *Spalding Official Baseball Guides* for 1938, 1939, and 1940 and for information on the Cape Breton Colliery League, the *Colliery League Digest* for 1939.

Correspondence and background material was provided by, among others, Colin Howell of Saint Mary's University in Nova Scotia, Donald F. Campbell of Sydney, Nova Scotia (son of Judge A.D. Campbell, president of the Cape Breton Colliery League), Charles Ballem of Dalhousie University in Halifax, the late Frank Graham, Honorary Secretary and Curator of the

Newfoundland Sports Archives, Gerard (Moe) Kiley who played for Glace Bay in the Cape Breton Colliery League, and the Over the Hill Dodger Gang: former members of the New Waterford team of the Colliery League.

Chapter 5
Western Canada

Books consulted included Hal G. Duncan's *Baseball in Manitoba* (Souris, Manitoba: Sanderson Printing, 1989), David Shury's *Play Ball Son! The Story of the Saskatchewan Baseball Association* (North Battleford, Saskatchewan: Turner Warwick Printers, 1986) and Shury's *Batter-Up: The Story of the Northern Saskatchewan Baseball League* (North Battleford, Sask.: Turner Warwick Printers, 1990). *Saskatchewan: A Pictorial History* (ed. Bocking, Western Producer Prairie Books) includes photos of early baseball in the province, Tony Cashman's *A Picture History of Alberta* (Edmonton: Hurtig, 1979) includes several photos of early baseball in Alberta; Tot Holmes' *Brooklyn's Babe: The Story of Babe Herman* (Gothenburg, Nebraska: Holmes, 1990), Barry Broadfoot's *Ten Lost Years 1929-1939* (Toronto: Doubleday, 1973), Barry Broadfoot's *The Pioneer Years 1895-1914* (Toronto: Doubleday, 1976), *Frontier Calgary 1875-1914* ed. Rasporich and Klassen (Calgary: University of Calgary, 1975), James H. Gray's *The Boy from Winnipeg* (Toronto: Macmillan, 1970) and particularly a chapter on the sporting scene in Winnipeg, Heather Robertson's *Salt of the Earth* (Toronto: Lorimer, 1974), *Saskatchewan Twilite Baseball* (ed. Andy Zwack, 1991), *The Saskatchewan Historical Baseball Review* published annually between 1984 and 1990 by the Saskatchewan Baseball Hall of Fame and Museum Association (ed. by David Shury) and the best source on baseball in western Canada.

Articles consulted include the various newsletters of the Saskatchewan Baseball Hall of Fame and Museum, David Shury's articles from his 'Prairie Diamonds' column in the *Battleford Express* 1982, Morris Mott's 'The First Pro Sports League on the Prairies: The Manitoba Baseball League of 1886' in the *Canadian Journal of History of Sport* (December 1984), Gary Bowie's 'The Beginnings of Sport in a Prairie Community' (a paper presented at the North American Society for Sport conference in 1980) on sports in Lethbridge.

Newspapers consulted included the *Edmonton Bulletin* for early baseball in the city, the *Regina Leader*

prior to World War One, the *Winnipeg Free Press*, and the *Winnipeg Daily Tribune*, the subject and biographical index of the *Lethbridge News* 1901-1906, and the *Lethbridge Herald* 1905-1918, and the *Lethbridge News and MacLeod Gazette* 1882-1900, excerpts from Western Canadian newspapers particularly the *Saskatoon Phoenix* on the Western Canada League 1908-14, story on a little league team in Calgary from the *Herald Magazine* 12 October 1973, story on minor league ball in Calgary from the *Herald Magazine* 21 July 1985, Jack Tennant's story on Russ Parker in the *Calgary Sunday Sun*, 17 August 1986.

Additional sources included programs and yearbooks of the Calgary Cannons and Edmonton Trappers baseball teams, various *Spalding Guides*, and the John Ducey Collection in the Provincial Archives of Alberta.

Correspondence and background materials provided by, among others, Wayne Vekteris of Calgary, Morris Mott of the University of Manitoba, Gary Bowie of the University of Lethbridge, Owen Ricker of Regina on 'The Strange Case of Dick Brookins', the late John Ducey (Edmonton's Mr Baseball), the late Babe Herman formerly of the Brooklyn Dodgers, and David Shury of North Battleford, Saskatchewan.

Chapter 6
British Columbia

Books consulted included Pat Adachi's *Asahi: A Legend in Baseball, a Legacy from the Japanese Canadian Baseball Team to its Heirs* (Toronto: Coronex, 1992) about Vancouver's best Japanese Canadian team, the Asahis, who played between the two World Wars, Harry Gregson's *A History of Victoria 1842-1970* (Victoria: Victoria Observer Publications, 1970), Seymour R. Church's *Baseball: The History, Statistics and Romance of the American National Game* (1974: Pyne Press republication of original work published in 1902), Harold Peterson's *The Man Who Invented Baseball* (New York: Scribner's, 1969) about Alexander Cartwright.

Articles consulted include several by Geoff LaCasse of Hippo Consulting and Research including 'Two Seasons to Remember: The Kamloops Baseball Championships, 1888 and 1889', 'The Development of Professional Baseball: Vancouver Before 1905', 'Early Professional Baseball in Vancouver: The

Northwestern League, Vancouver Beavers, and R.P. Brown', plus an article 'From Amity Wolf to Vancouver Beaver' on British Columbia baseball in *Dugout* magazine (June/July 1994), George Bowering's review of 'Cheering for the Home Team' in *Brick* magazine (Spring 1984), George Bowering's stories in *Descant* 56/57 (Spring-Summer 1987), Lorne W. Rae's 'It Was Real Baseball' from *Saskatchewan History* XLIII, 1 (Winter 1991) on the 1935 visit of a Japanese all-star team to Saskatoon.

Newspaper sources consulted included the *Whitehorse Star* and the *Yukon Sun* of Dawson City for the early 20th-century period.

Additional printed materials include Vancouver Mounties and Vancouver Canadians programs.

Correspondence and additional materials were provided by, among others, Geoff LaCasse on Victoria's first team the Amity, Dave MacLean on B.C. Place, Pat Adachi, Ed Kitigawa and Tom Matoba on the Asahis, and Bill Fitsell on Robert Eilbeck.

Chapter 7
Québec

Books consulted included Merritt Clifton's *Disorganized Baseball: The Provincial League from La Roque to Les Expos* (Richford, Vermont: Samisdat, 1982), Donald Guay's *Introduction à l'histoire des sports au Québec* (Montreal: VLB Editeur, 1987) and the chapter 'Le baseball'; Jean-Marc Paradis' *100 ans de baseball à Trois-Rivières* (Trois-Rivières, 1989) and his *Histoire illustrée du baseball rural en Maurice, 1940-1990* (Trois-Rivières, 1990), Jean Blouin's *Roland Beaupre, Monsieur Baseball, se raconte* (Montreal: Les Presses Libres, 1980), Dan Turner's *The Expos Inside Out* (Toronto: McClelland and Stewart, 1983), Roger Lemelin's *The Plouffe Family* (Toronto: McClelland and Stewart, 1950), Horace Miner's *St Denis: A French-Canadian Parish* (Chicago: University of Chicago Press, 1939), Rioux and Martin's *French-Canadian Society* Vol. 1 (Toronto: McClelland and Stewart, 1964), Everett Hughes' *French Canada in Transition* (Chicago: University of Chicago Press, 1943), Brodie Snyder's *The Year the Expos Almost Won the Pennant* (Toronto: Virgo Press, 1979), Brodie Snyder's *The Year the Expos Finally Won Something!* (Toronto: Checkmark, 1981).

Articles consulted included Mordecai Richler's 'Kermit Kitman Played Here' from the *Weekend Magazine* (7 April 1979), Robert Fontaine's 'God Hit a Home Run' which appeared in *The Second Fireside Book of Baseball* (ed. Charles Einstein, 1958) and describes a Sunday baseball game in Hull, Quebec; Alan Metcalfe's 'Organized Sport and Social Stratification in Montreal: 1840-1901' from *Canadian Sport: Sociological Perspectives* (ed. Gruneau and Albinson, 1976), Peter Bjarkman's 'Montreal Expos: Bizarre New Diamond Traditions North of the Border' from *Encyclopedia of Baseball Team Histories; National League* (ed. Bjarkman, 1993), Paul Post's 'Origins of the Montreal Expos' from *The Baseball Research Journal* of the Society for American Baseball Research (No. 22, 1993).

Newspapers consulted included the Montreal *Gazette* particularly for August 1869 and for 19th-century baseball, *Le Soleil de Quebec*, the Toronto *Globe and Mail* for the 1946 International League season, Gord Walker of the *Globe and Mail* on major league baseball's 1969 debut in Montreal, a 1982 story by Montreal sportswriter Dink Carroll on Montreal's baseball tradition.

Other written records consulted included a story by E.J. Blandford on the Dow Breweries' sports sponsorship, the *Montreal City Directory*, and various programs of the Montreal Royals and Montreal Expos.

Correspondence and background materials were provided by, among others, Merritt Clifton, Bobby Fisher of the Sherbrooke *Record*, Dan Ziniuk on nationalism and baseball in Quebec, Irwin Pinsky on the Plouffe Family, and Michel Vigneault on Quebec sports sources.

Chapter 8
Baseball in One Canadian Community

Books consulted included Hall and McCulloch's *Sixty Years of Canadian Cricket* (Toronto: Bryant Printing, 1895), *The Canadian Cricketer's Guide* (1858), *The Canadian Cricket Guide 1876* (Ottawa Free Press), Joseph Schull's *Edward Blake* (Toronto: Macmillan, 1975), *Directory of the County of Northumberland and the County of Durham* (Port Hope: E.E. Dodds, 1880), John Squair's *The Townships of Darlington and Clarke* (Toronto: University of Toronto Press, 1927), ed. Garfield Shaw et al., *Picture The Way We Were: A Pictorial History of Darlington and Clarke Townships* (1980), James B. Fairbairn's *History and Reminiscences* of *Bowmanville* (Bowmanville: Bowmanville News Print, 1906), Hamlyn, Lunney, and Morrison's *Bowmanville: A Retrospect* (Bowmanville: Bowmanville Centennial Committee, 1958).

Newspapers and magazines consulted included the *Canadian Statesman* in the Bowmanville Library, *The Globe* (later *The Globe and Mail*) (16 July 1855 and other issues), *The Belvedere*, a quarterly journal of the Bowmanville Museum. Other written records consulted included 'History of St John's Anglican Church' which appeared periodically in 1935 in the *Canadian Statesman*.

Chapter 9
Black Baseball in Canada

Books consulted included Robert Peterson's *Only the Ball was White* (Englewood Cliffs, N.J.: Prentice-Hall, 1970), Donn Rogosin's *Invisible Men: Life in Baseball's Negro Leagues* (New York: Atheneum, 1985), James A. Riley's *The Biographical Encyclopedia of the Negro Baseball Leagues* (New York: Carroll and Graf, 1994), John Craig's *Chappie and Me* (New York: Dodd, Mead, 1979), Clark and Lester, eds, *The Negro Leagues Book* (Cleveland: SABR, 1994).

Articles consulted included E.G. Hastings' 'Baseball once the big sport in Dunnville' from the *Dunnville Chronicle* (31 August 1977).

Correspondence and background materials were provided by, among others, L. Robert Davids on Alex Ross.

Chapter 10
Owners, Organizers, and Players in Ontario Baseball

Books consulted included Louis Cauz's *Baseball's Back in Town: A History of Baseball in Toronto* (Toronto: CMC, 1977), David Pietrusza's *Baseball's Canadian-American League* (Jefferson, North Carolina: McFarland, 1990), Tom Anderson's *Memories: A History of the North Dufferin Baseball League* (self-published, 1980), George Vass's *Like Nobody Else: The Fergie Jenkins Story* (Chicago: Henry Regnery, 1973), Jesse Middleton's *Toronto's 100 Years: 1834-1934* (Toronto: Centennial Committee, 1934), Marchildon and Kendall's *Ace: Canada's Pitching Sensation and Wartime Hero* (Toronto: Viking, 1993) on Phil Marchildon.

Articles consulted included Stephen Gamester's 'You can't tell the Canadian Big League Heroes with-

out a program: Here it is' from *Maclean's* (22 August 1964), Paul Patton's 'Canadian ties strong in history of baseball', *Globe and Mail* (24 June 1985), Paul Patton's 'Emslie was first—by 83 years', the *Globe and Mail* (9 July 1974) on Canadian-born major league umpire Bob Emslie.

Other written records consulted included those in the Bob Emslie collection in the Elgin County Museum in St Thomas, Ontario.

Chapter 11
Toronto Blue Jays

Books consulted included Rosie DiManno's *Glory Jays: Canada's World Series Champions* (Toronto: Sagamore, 1993), Peter Bjarkman's *The Toronto Blue Jays* (Toronto: B. Mitchell, W.H. Smith, 1990), Jon Caulfield's *Jays! A Fan's Diary* (Toronto: McClelland and Stewart, 1985), Larry Grossman's *A Baseball Addict's Diary: The Blue Jays' 1991 Rollercoaster* (Toronto: Penguin, 1992), Larry Millson's *Ballpark Figures—The Blue Jays and the Business of Baseball* (Toronto: McClelland and Stewart, 1987), Van Rjndt and Blednick's *Fungo Blues—An Uncontrolled Look at the Toronto Blue Jays* (Toronto: McClelland and Stewart, 1985), Alison Gordon's *Foul Balls! Five Years in the American League* (New York: Dodd, Mead, 1984), Buck Martinez's *From Worst to First—The Toronto Blue Jays in 1985* (Toronto: Fitzhenry and Whiteside, 1985), Whitt and Cable's *Catch—A Major League Life* (Toronto: McGraw-Hill Ryerson, 1989) on Ernie Whitt, Stuart Broomer's *Paul Molitor: Good Timing* (Toronto: ECW Press, 1994), David Fulk, ed., *A Blue Jays Companion* (South Pasadena, CA: Keystone, 1994).

Articles consulted included Peter Bjarkman's 'Toronto Blue Jays: Okay Blue Jays! From Worst to First in a Decade' in *Encyclopedia of Major League Baseball Team Histories Vol. 1: American League* (New York: Carroll and Graf, 1993)

Newspapers consulted included *USA Today: Baseball Weekly* for the 1992 and 1993 season.

Appendix A
Canadian Big League Players

Primary sources included *Total Baseball* (listed above), *The Baseball Encyclopedia* (listed above), lists from a variety of sources including *The Canadian Sportscard Collector* (December 1993), members of the Society for American Baseball Research including Bill Carle, Peter Morris, Peter Bjarkman, and Eves Raja.

Appendix B
Canadian Women in the AAGBL

Primary sources included the Baseball Hall of Fame in Cooperstown, New York, and the AAGBL Cards (Box 3332, Kalamazoo MI, 49003-3332) produced by Sharon Roepke. Books consulted included Barbara Gregorich's *Women at Play: The Story of Women in Baseball* (San Diego: Harcourt, Brace, 1993), Susan E. Johnson's *When Women Played Hardball* (Seattle: Seal Press, 1994), Lois Browne's *Girls of Summer: In Their Own League* (Toronto: Harper Collins, 1992), Sharon Roepke's *Diamond Gals: The Story of the All American Girls Professional Baseball League* (Marcellus, Michigan: AAGBL Cards, 1986).

Appendix C
Canadian Teams in Organized Baseball

Primary sources included *Professional Baseball Franchises* (listed above), *The Encyclopedia of Minor League Baseball* (listed above), Robert Objoski's *Bush League: A History of Minor League Baseball* (New York: Macmillan, 1975), the Spalding and the Reach official guides at the Baseball Hall of Fame and Museum in Cooperstown, New York.

General Sources

Books consulted included William Humber's *Cheering for the Home Team* (Erin, Ontario: Boston Mills Press, 1983), William Humber's *Let's Play Ball:* *Inside the Perfect Game* (Toronto: Lester and Orpen Dennys/Royal Ontario Museum, 1989), Dan Turner's *Heroes, Bums and Ordinary Men: Profiles in Canadian Baseball* (Toronto: Doubleday, 1988), Jim Shearon's *Canada's Baseball Legends: True Stories of Canadians in the Big Leagues since 1879* (Kanata: Malin Head Press, 1994, Gruneau and Whitson's *Hockey Night in Canada: Sport Identities and Cultural Politics* (Toronto: Garamond Press, 1993), W.A. Hewitt's *Down the Stretch: Recollections of a Pioneer Sportsman and Journalist* (Toronto: Ryerson Press, 1958), Henry Roxborough's *One Hundred—Not Out: The Story of 19th Century Canadian Sport* (Toronto: Ryerson, 1966), Brenda Zeman's *To Run With Longboat: Twelve Stories of Indian Athletes in Canada* (Edmonton: GMS Ventures, 1988) particularly the chapter on 'The Smilin' Rattler': Jimmy Rattlesnake', Bob Ferguson's *Who's Who in Canadian Sport* (Toronto: Summerhill Press, 1985), John Bell, ed. *The Grand Slam Book of Canadian Baseball Writing* (Lawrencetown Beach, Nova Scotia: Pottersfield Press, 1994), Hall, Slack, Smith, and Whitson's *Sport in Canadian Society* (Toronto: McClelland and Stewart, 1991).

Magazines consulted included *Baseball America* founded by a Canadian Alan Simpson in White Rock, British Columbia in 1980. Simpson later moved to Durham, North Carolina as the magazine's editor after Miles Wolff purchased it in 1982. Others included *USA Today—Baseball Weekly*, and three Canadian baseball publications, *Bullpen: The Canadian Baseball Review* (1980), *Innings* (mid 1980s, publisher Martin Levin), and *Dugout* (founded 1993, publisher Russell Field).

INDEX

Italicized page numbers: photos

Aaron, Hank, 148
Abbott, Jim, 174
Adams, Daniel, 23
Adams, Frank, 159
Allen, Ebenezer, 17
Allen, G.G., 57
Alomar, Roberto (Robbie), 3, 172, 173, 176, 177, *178*, 179, 180
Alou, Moises, 124
Alston, Walter, 107
Ames, Fisher, 21
Amoros, Sandy, 117, *148*
Anderson, James Robert, 19
Anderson, Nuts, 89
Anderson, Sparky, 176
Angell, Roger, 13, 169
Anson, Cap, 29, 39
Archer, John, 75
Armour, Robert, 133
Atkinson, Eliott, 49
Averill, Earl, 161

Babb-Sprague, Kristen, 177
Babe, Loren, *166*
Babe, W., 89
Bachiu, Larry, *11*
Bailey, Nat, 102, 103, 104
Bailor, Bob, 175
Baker, Mary 'Bonnie', 79–80, 162
Bailey, John, 131
Ballem, Charles, 54

Bankhead, Dan, 117
Bankhead, Sam, 121
Banta, Jack, 115
Barbados, 147, 163
Barber, George Anthony, 130, 131
Barfield, Jesse, *174*
Barnes, Frank, 166
Barnes, Roscoe, 39
Barney, Robert (Bob), 1, 18, 114
Barnfather, J.H., 10, 85
Barrette, Jean, 118
Barrow, Ed, 113, 153
Baseball (base ball, Base Ball), 1, 3, 6, 15, 19, 134; Alberta Major Baseball League, 73; All-American Girls Professional Baseball League (AAGBL), 11, 75, 79, 80, 71, 162; amateur ball in Vancouver, 99–102; amateur Terminal League, 99; American Association, 50, 84, 143, 156; American League, 7, 13, 74, 119, 122, 125, 126, 167, 169, 173, 176, 177, 180, 181; armed forces teams, 10; Bantam, 13, 14; *Baseball Encyclopedia*, 171; Big Four semi-pro league, 71; Border League, 118, 120, 157, 164; British Columbia Amateur League, 98–9; buckboard league, 90; Canadian Amateur Baseball Association, 12, 48; Canadian-American League, 120, 57, 161; Canadian Association of Ballplayers, 32, 34; Canadian Association of Base Ball Players, 4, 5, 12; *Canadian Base Ball Guide*, 5; Canadian Baseball Hall of Fame, 90, 91, 106, 107, 158; Canadian Federation of Amateur Baseball (CFAB), 12, 13; Canadian League, 7, 135, 151, 157, 159; Canadian National Beaver championship, 165;

Canadian Pan-American Games team, 11, 12; Central Ontario League, 157; Class 'A' Pioneer League, 73, 74; Class 'A' Western International League, 72; Class 'B' Eastern Canada League, 118; Class 'C', 50; Class 'D' league, 48–9, 50; Colliery League, 48, 50; Colored All-Stars, 118, 147; Colored Giants, 87; Continental League, 155; Cy Young award, 163; diamond of 1876, 24; Eastern International League, 104; Eastern League, 7, 31, 73, 156, 157; emery ball, 90; Federal League (US), 9, 37, 92; Gold Medal, amateur, 14; Halifax and District Summer League, 162; House of David team, 118; International Association, 7, 33, 34, 36, 37, 39, 136, 159; International League, 7, 50, 107, 111, 112, 113, 114, 115, 116, 117, 119, 120, 143, 150, 154, 155, 156, 166, 172; Interprovincial play, 42; junior, 13; Junior World Series, 113, 114, 115, 117, 120; 'la balle au camp', 108; Little League, 13, 14; Little World Series, 50, 55, 114; Lou Marsh trophy, 163; Mandak League, 88, 89; Mantitoba League, 85, 88; Manitoba-Saskatchewan League, 87; Maritime Senior championship, 52; Massachusetts Game (townball), 18, 19, 24, 42; Mexican League, 121; Michigan-Ontario League, 157; Midget, 13; Minor Leagues, 6, 74–5; Mosquito, 13; National Agreement, 7; National Association, 29; National Association of Professional Baseball Leagues, 7, 9, 48, 50; National Baseball Hall of Fame and Museum (US), 2, 25, 56, 163, 165; National Board, 7; National Commission, 67; National League (US), 6, 7, 12, 32, 34, 35, 36, 37, 38, 39, 50, 84, 86, 113, 123, 124, 125, 161, 165, 169, 181; Native baseball in Canada, 59–61; Negro Leagues, 122, 139, 146, 147, 149; New England League, 157; New York rules, 24, 25, 27, 28, 42, 77; Nickel Belt League, 162; Nippon teams, 99; North-Central League, 181; Northern League, 89; Old Swamp Angel team, 236; Ontario Baseball Association, 9, 144; Ontario Inter-City League, 165; Ontario League, 157; 'Organized Baseball', 7; *Pacific Base Ball Guide*, 93; Pacific Coast League (PCL), 72, 73, 115, 144, 161; Pearson Cup, 167; Pee Wee, 13; Player Development Contract, 73; pop fly, 29; Professional Northwestern League, 99; Provincial League/La Ligue Provinciale, 117, 118; Quebec-Ontario-Vermont League, 118; Quebec Senior League, 106; rabbit ball, 30;

regional variations, 18; Royal Ontario Museum, 2; Saskatchewan Baseball Hall of Fame, 79, 82; semi-pro leagues, 118, 121, 122; Silver Ball (competition), 4; slider pitch, 160; sliding technique, 171; Society for American Baseball Research, 2; Jonathon League, 180; spitball throw, 157; Three-1 League, 90; townball, 18, 19; Union League, 37; Western Canada League, 9; Western Canada B League, 89; Western League (US), 92; Western Ontario League, 7; Western Ontario B League, 152; women's teams, 43, 45, 60, 77; World Semi-Professional Championship, 30; World Series, 1, 2, 13, 19, 28, 39, 51, 58, 117, 123, 124, 125, 126, 161, 165, 168–77; World Youth Championship, 14
basketball, 37, 123, 164
Bearnarth, Larry, 179
Beaver, Bev, 60
Beaver, George, 60
Beers, George, 8, 109
Beeston, Paul, 168, 169, 173, 174
Belcourt, A., 76
Bell, Alexander Graham, 48
Bell, Derek, 174, 179
Bell, George, 173, 174, *175*, 178
Bender, Chief Albert, 91
Benedicts, 127
Bennett, Captain, 105
Berg, Moe, 118
Bermuda, 92
Bertoia, Reno, 158, 159, 162
Betzel, Bruno, *166*
Biasatti, Hank, 158, 159
bicycling, 63
Bishop, Hub, 80, 81
Bissonette, Del, 49, 109, 118, 120
Black, Joe, 117
Blackburn family, 38
Blake, Eddie, *166*
Blake, Edward, 130
Blake, Toe, 118
Blyleven, Bert, 72
Boehm, Bob, 80
Bolger, Dermot, *The Tramway End,* 168, 169
Bolton, US consul, 98
Bookless, William, 28
Boone, Ike, *154*
Borders, Pat, 172, 179
Boucher, Denis (1), 106, 125

Boucher, Denis (2), 107
Bouchier, Nancy, 1, 18
Bousquet, Joseph, 118
Boutlier, Martin, 165
Bouton, Jim, 13
Bowering, George, 96
Bowman, Charles, 127
Bramham, Judge, 50
Branca, Ralph, 117
Brannock, Mike, 29
Bransfield, Kitty, 113
Breard, Roger, 121
Breard, Stan, 114, 120, 121
Brett, George, 161
Brewer, Chet, 83
Brigham Young, J., 85, 86
Broadbent and Overall, 38
Bronfman, Charles, *122*, 124
Bronson, Les, 9
Brookins, Dick, 9, 67
Brother Matthias, 165
Brown, George, 21, 130
Brown, Joe, *100*
Brown, John, 32, *33*
Brown, Bob 'Ruby Robert', 102, 103
Brown, Tom, 107
Brunelle, Leo, *76*
Bryce, William, *Canadian Base Ball Guide,* 6, 24
Burden, Samuel, 127, 133, 135
Burley, Mr, 135
Burns, Thomas, Collection, 85
Burroughs, Jeff, 164
Burtch, Zachariah, 25
Bush, Carmen, 157
Butler, Frank, 58
Butler, Rich, 58
Butler, Rob, 58–9, 179

Cabana, Homer, 118
Cairns, J.F., 69
Caisse, Wilfrid, 108
Calfron, E., *76*
Callaghan, Helen, *81*
Callaghan, Margaret, *81*
Callaghan, Morley, 159
Calvert, Paul 'Cyclops', 118, 119
Campanella, Roy, 115
Campanis, Al, 115
Campbell, Judge Andrew Dominic (A.D.), 49
Campbell, Clarence, 171

Campbell, Lou, 52, 53
Canada: Amateur Athletic Association, 109; Amateur Athletic Union, 11; Baseball Hall of Fame, 117; Football Association, 109; Olympic Association, 109; Olympic Games team, 58–9; Wheelmen's Association, 109
Alberta, 4, 75, 141; Alix Fair, 64; Amber Valley (Pine Creek), 144, 145; black team, 61; Calgary, 1, 62, 63, 64, 65, 66, 67, 70, 72, 73, 81, 96, 144; Cereal, 68; Edmonton, 8, 62, 65, 67, 69, 70, 71, 72–3, 75, 144, 178; Edson, 69; Fort McMurray, 14; Hobbema Reserve, 59; Jasper, 69; Lethbridge, 62, 63; Medicine Hat, 62, 65, 96, 159; Slave Lake, 61; Wetaskiwin, 67
British Columbia, 4, 13, 75, 93–105, 141, 144, 175; Donald, 95, 96; Fraser Valley, 19, 99; Kamloops, 63, 93, 95, 96, 98, 149; Kaslo, 95; Lake Windermere, 97; Nanaimo, 93, 98; New Westminster, 65, 93, 94, 96, 98, 111; Nicola, 95; Ocean Falls, 99; Okanagan, 99; Steveston Fujis team, 104; Vancouver, 3, 9, 72, 95, 96, 98–104, 165; Vancouver Island, 19; Victoria, 4, 19, 72, 93, 94, 95, 96, 97, 98, 99, 160; Wellington, 144
Manitoba, 12, 75, 77, 82–92, 176; Assiniboia River, 19; Brandon, 14, 67, 87, 88–9; Carman, 87, 89; Crandall, 87; Dennis County, 87; Garry, 83; Hamiota, 87; Indian Head, 89; Miniota, 87; Oak Lake, 87; Portage La Prairie, 83, 84, 85, 89; Queen Valley, 87; Red River, 19, 20, 83; Souris, 87; Virden, 87, 89; Waskada, 87; Winnipeg, 8, 9, 10, 12, 67, 71, 75, 82, 83, 84, 85, 89, 172
New Brunswick, 9, 44, 45, 141; Black's Harbour, 47; Bocabec, Charlotte County, 43; Fredericton, 44, 144; McAdam, 53; Moncton, 44, 47, 53; Perth, 44; Saint John, 8, 22, 41, 42–5, 46, 47–8, 50, 53, 54, 144
Newfoundland, 3, 22, 55–9
Northwest Territories, 3, 20, 63
Nova Scotia, 11, 12, 141, 165; Acadian teams, 59; Amherst, 53; Annapolis Valley, 165; Cape Breton, 12, 48–50, 59–60; Dartmouth, 21–2, 45, 54, 144; Glace Bay, 48, 49, 50; Halifax, 4, 8, 9, 12, 21–2, 41, 45–7, 49, 50, 51, 53, 109, 144, 145; Joggins, 48; Lingan, 165; Liverpool, 165; Maliseet people, 22; Mi'kmaq people, 22;

Milton, 90; New Glasgow, 52; New Waterford, 49, 50; Penobscot people, 22; Pictou, 48, 51; post-war senior baseball, 54–5; Reserve Mines, 48, 49; Springhill, 47, 48, 50, 52, 53; Stellarton, 54; Sydney, 47, 48, 49, 50, 54, 165; Truro, 144, 162; Westville, 48, 53 165; Yarmouth, 50, 53, 165

Ontario, 5, 12, 18, 26, 75, 109, 150; Amherstburg, 158; Barrie, 153, 162; Beachville, 1, 16, 17, 18, 20, 22; Belleville, 133; Bowmanville, 2, 14, 34, 127–38, 165, 175; Brantford, 29, 60, 146, 148, 149, 157, 165; Brewers and Malters Association, 37; Canadian League, 62; Caughnawaga, 59; Chatham, 141, 147, 157, 163, 164, 165; Clarke Township, 135; Cobourg, 132, 133; Creighton Mines, 162; Cumberland township, 14; Darlington, 130–2, 135; Delaware, 17; Dundas, 27, 29; Durham, 131, 134; Elmvale, 162; Elora, 29; Emily, township of, 9; Englehart, 37; Enniskillen, 134; Fergus, 29; Fort William, 160; Gloucester, 14; Goderich, 38; Gore Bay, 162; Guelph, 1, 5, 7, 28, 29, 30, 31, 36, 37, 89, 111, 143, 144, 150, 151, 161, 165, 179: Maple Leafs, 28–32, 33, 34, 35, 39, 134, 135, 136, 142; Hamilton, 4, 7, 8, 10, 16, 19, 23, 24, 25, 26, 27, 32, 34, 36, 39, 40, 46, 85, 111, 133, 135, 136, 143, 144, 150, 157: Maple Leafs, 19, 25, 26, 28; Standards, 32, 136; Harriston, 138; Ingersoll, 4, 5, 6, 9, 17, 25, 27, 28, 152, 162; Inter-County Major League, 146, 148, 149; Kapuskasing, 72; Kingston, 4, 5, 32, 104, 109, 157, 159: St Lawrence team, 104, 134, 135; Kitchener, 157; Lindsay, 133; London, 1, 5, 6, 8, 17, 19, 20, 24, 26, 29, 32, 33, 34, 36, 37, 38, 39, 40, 111, 135, 136, 155, 157, 159: Atlantics, 39, 136, Baseball Club, 32, 33, Goodwills black team, 142, Majors, 165; Tecumsehs, 6, 16, 32, 33, 34, 35, 37, 38, 39, 40, 135, 136, 142, 159; Markdale, 89; Markham, 136; Midland, 162; Napanee, 133; Newburgh, 159; Newcastle, 132, 133, 134, 135; Newmarket, 15; Niagara, 17, 23; North Oxford, 17; Northumberland-Durham, county of, 132; Oakville, 144; Orillia, 162; Orleans, 14; Orono, 127, 135; Oshawa, 133, 149; Ottawa, 4, 6, 8, 14, 107,

116, 139, 162, 164, 172; Penetang, 162; Peterborough, 157; Petrolia, 142; Port Hope, 133, 135, 175; Port McNicoll, 162; Port Perry, 133; Port Stanley, 38, 159; Prescott, 27; Ridgetown, 159, 160; St Catharines, 141; St David's, 30; St Marys, 16; St Thomas, 7, 144, 157; Sarnia, 89, 179; Seventh Line, 134–5; Simcoe County, 130; Smiths Falls, 157; Stratford, 179; Toronto, 1, 2, 7, 8, 9, 10, 12, 15, 16, 21, 24, 25, 26, 28, 72, 111, 122, 124, 130, 132, 133, 134, 135, 136, 139, 141, 143, 144, 150: Blue Jays, 1, 3, 13, 14, 19, 26, 39, 58–9, 61, 72, 73, 82, 106, 123, 165, 168–82, International League team, 48, Maple Leafs, 12, 119, 151, 154, 155, 156, 157, 162, 166, Oslers, 89, Parkdale Baseball Club, 150, Sunnyside, 11, University of Toronto, 132, 159, Upper Canada College, 16, 130, 131; Whitby, 131, 133; Windsor, 17, 19, 107, 141, 158–9, 160; Woodstock, 4, 17, 20, 22, 25, 26, 27, 28, 29, 31, 89, 113, 132, 136, 141, 142, 160; York, 15; Zorra Township, 16, 17

Prince Edward Island, 13, 50–4, 59–60

Quebec, 8, 12, 13, 18, 20–21, 106–26, 156; Aces hockey team, 54; Beauce, 109; Cantons de l'Estrie, 21; Cap Madeleine, 109; Caughnawaga, 59, 118; Causapscal, 121; Chateauguay Basin, 21; Dow Brewery Nine, 118; Drummondville, 118, 120, 121; Eastern Townships Industrial League, 118, 120; Farnham, 112, 118, 121; Gaspé, 109; Granby, 118, 119, 122; Hull, 112; Huntington, 21, 22; Ile d'Orléans, 139; Lachine, 118, 120; Lévis, 106; Ligue Commerciale, 120; Lique Provinciale, 112, 117; Louiseville, 108; Montreal, 1, 8, 12, 13, 14, 21, 41, 107, 109–17, 118, 119–20, 123, 125, 132, 139: Amateur Athletic Association (MAAA), 51, 109, Expos, 3, 6, 12, 13, 14, 54–5, 73, 106, 107, 112, 122–5, 165, 179, 181, McGill University, 18, 31, 'O.K. Bill', 113, Royals, 113, 114, 116, 117, 121, 148; Quebec Provincial League, 117–22; Quebec City, 41, 118, 120, 165; Richelieu League, 120; Rimouski, 121; Saranac Lake, 118; Sherbrooke, 118, 119, 120, 121; Sorel, 112, 118; St-Denis, 107; Ste-Anne-de-Bellevue, 108, 118; St-Hyacinthe, 3, 107, 108, 112, 118, 120; St-

Jean 112, 121; St-Jean DeMatha, 108; St-Mathias, 108; St-Maurice Paper, 109; Trois-Rivières, 3, 108, 109, 119; Valleyfield, 112; Victoriaville, 118; Yamaska League, 120

Saskatchewan, 4, 11, 75–82, 86, 87, 149, 176; Broadview, 75; Carrot River Loggers, 89; Colliston, 75; Fort Battleford, 75; La Fleche, 76, 80; Lumsden, 75; Melaval, 80; Melville, 82; Moose Jaw, 67, 69, 77, 78; Moosomin, 73, 87; Nipawin, 82; North Battleford, 12, 82, 149, 165; Ogewa, 80; Prince Albert, 75; Qu'Appelle Valley, 75, 76; Regina, 67, 69, 75, 77, 79, 80, 81; Saskatoon, 3, 12, 67, 68, 69, 102, 149, 181; Snowden, 82; Unity, 149; Watrous, 82

Yukon, 94, 104–5

Canadian Broadcasting Corporation (CBC), 12

Canadian Federation of Amateur Baseball (CFAB), 12–14

Canadian Imperial Bank of Commerce (CIBC), 169

Canadian Pacific Railway (CPR), 44, 65, 75, 83, 84, 85, 86, 93, 95, 96, 98, 111, 118, 147

Cauz, Lou, 151

Carleton, Sir Guy, 17

Carnegie, Herbie and Ossie, 139

Caroll, Captain John, 17

Carter, Gary, 123, 124

Carter, Joe, 1, 3, 14, 169, 171, 172, 173, 176, 177, 178, 179, 181

Cartwright, Alexander, 13, 18, 93, 94

Cartwright, Alfred, 93, 94

Case, Orrin Robinson 'Bob', 159

Casey at the Bat, 169

'cat', 15

Chaplin, Mark, 57

Chase, Hal, 91, 97

Cheek, Tom, 174

Chesterfield, Lord, 92

China, 98, 102

Choquette, Joseph, 118

Christensen, Walter 'Cuckoo', 70

Clancy, 'Hustlin' Irish', 48

Clarke, George, *The Best One Thing,* 9

Clarke, Jay 'Nig', 158

Claxton, Jimmy, 142, 144

Cleveland, President Grover, 90

Cleveland, Reggie, 82

Clifford, Luther 'Shanty', *147,* 148

Clifton, Merritt, 117, 118, 120

Climie, George, 136

Climie, John, 130

Climie, William Roaf, 127, 130, 132, 133, 134–5, 136, 165

Clyde, 25, 26

Coakley, Andy, 45

Cobb, Ty, 92, 102

Coderre, Shorty, 118

Cogswell, Dr A.C., 45

Coleman, Edwin, 136

Colson, Jim, 28

Colwell, Ira, *65*

Conan Doyle, Sir Arthur, 69

Cone, David, 171, 172, 176, 177

Conigliaro, Tony, 55

Connor, Ralph, *The Sky Pilot: A Tale of the Foothills,* 62

Connors, Chuck, 115

Connors, 'Putty', 53

Coombs, Jack 'Cy', 45

Cooper, Jean (Powless), *60*

Corey, Gene, *11*

Cormier, Rheal, 55

Corrales, Pat, 164

Cory, Cliff, 89

Cottingham, Sherman, 149

Courtroille, Joe, 61

Cowan, Joe, 89

Cox, Chesty, 67

Cox, E. Strachan, 151

Craig, John, 120

Craig, Pete, 158

Craig, Roger, 82

Craver, Bill, 37

Crawford, 'Wahoo Sam', *157*

cricket, 3, 5, 7, 9, 10, 15, 19, 27, 32, 56, 83, 94, 95, 104, 109, 130–2, 135, 138, 165

Crimian, Jack, *166*

Cronin, Joe, 126

Cronin, Tom, 126

Cuba, 13, 89, 117, 127, 130, 133

Cubitt, Frederick, 127, 130, 133, 165

Cubitt, Richard, 129, 130

Cunningham, Cornelius, 17

curling, 127

Dadeaux, Rod, 72

Dash, 25

Davey, John, 111

David, Athanase, 113

Davids, L. Robert, 143

Davies, David, 27

Davis, Cal, 7
Davis, Geena, 80
Davis, Gladys 'Terrie', 80, 162
Davis, Premier William (Bill), 167
Davison, Rob, *46*
Dawson, Andre, 102, 123, 172
Delage, Emilion, 108
Delaney, Patrick, 118
Demarais, Fred, 108
DeMars, Billy, *166*
De Montreville, Eugene Napoleon, 108
Dempsey, Rick, 82
De Shields, Delino, 124
Desjardins, Marcel, 123
Devine, Lulu, 172
Devlin, Jim, 39
diamond of 1876, 24
Diggs, Fred, 144
DiMaggio, Joe, 120, 161
Dinnen, *33*
DiRubio, Johnny, 49
Dodge, William, 16
Doerksen, Irv, *11*
Dominican Republic, 106
Donahue, Terry, 80
Dooley, Charlie, 112
Dorion, Phil, *11*
Douglas, James, 19
Downey, Ellen, 136
Doyle, 'Dizzy', 48
Doyle, James, *46*
Drabowsky, Moe, 54
Dryden, M.J., 48
Drysdale, Don, *117*
Ducey, John, 71, 72
Ducey, Rob, 174
Dudley, US consul, 98
Dufferin, the Earl of, 32
Duffy, Lizzie, 43
Dugas, Gus, 108
Duncan, Hal, *Baseball in Manitoba*, 87, 89, 92
Dundas, George, 89
Durocher, Leo, 109, 164
Dussault, Normand, *119*, 120
Dyer, C.J. 33
Dygert, Warner, 17
Dynamite Lady, 74

Eaton, John Craig, 159
Eckersley, Dennis, 176, 177

Eckstrom, President C.J., 67
Edwards, Archie, 88
Egan, Lethbridge shortstop, 65
Eichhorn, Mark, 174
Eilbeck, J.M., 105
Eilbeck, Robert J., 104
Ejima, Frank, 102
Elias, John, *11*
emery ball, 90
Emery, Jacques, 108
Emslie, Bob, 136, 138
England, 19, 21, 22, 26, 27, 46; Erpingham, Norfolkshire, 127; Leeds, 79; London, 62
Englehart, Jacob, 37, 39
Europe, 20, 69, 156, 159
Ewasiuk, Ken, *11*

Fairbairn, J.B., 127, 130
Fairly, Ron, 72
Fanning, Jim, 55
Farquhar, James, *46*
Farrington, Harvey, 25
Feast, Arthur, 25, 28
'feeder', 15, 20
Felix, A., 111
Felsch, Hap, 87, *88*
Ferguson, Philip 'Skit', 54
Fernandez, Chico, *148*
Fernandez, Tony, *173*, 179, 180
Field, Russell, *166*
Fields, Wilmer, 148
Fingers, Rollie, 55
Finnamore, Art, *46*
Fisher, Traves, 89
Fisk, Carlton, 13, 172
Fleming, Dave, 74
Flood, Brian, 22, 42, 43
Flynn, Jocko, *46*
Fontaine, Robert, *The Happy Time*, 112
Foley, Ace, 50
football, 4, 12, 37, 39, 113, 115, 123, 139, 155
Foote Collection, 86
Forard, Hervé, 108
Ford, Dr Adam, 16, 17, 18, 20
Ford, Gene, 90
Ford, Philadelphia Phillies player, 164
Ford, Ida, 90
Ford, Russell (Russ), 89–92, *90*, *91*
Ford, Walter, 89–90
Forget, *Histoire du Collège de l'Assomption*, 108

Fowler, Bud, 142, 143
Fowler, Dick, 161
Frawley, 76
Frias, Pepe, 55
Furumoto, Ted, 100

Galarraga, Andres, 73
Gallagher, catcher, 96
Gallivan, Danny, 49
Galloway, William (Bill 'Hippo'/'Hippie'), 141, 144
Galt, Sir Alexander, 62
Galt, Elliott, 62
Ganong, Whidden, 50
Garcia, Damaso, 171
Gardella, 120, 121
Gaston, Cito, 169, 173, 174, 175, 179, 180, 181
Gauthier, Joseph, 108
Gauvreau, Colonel Romeo, 113
Geier (Giguere), Philip, 108
Gelinas, Arthur, 108
German (Germain), Les, 108
Gibbs, Scotty, 121
Gibson, Bob, 12
Gibson, Captain, 17
Gibson, Joseph, 6, 27
Gibson, Josh Jr, 121
Giles, Warren, 123
Gillean, Tom, 6, 33, 38
Gilliam, Junior, 117
Gillick, Pat, 72, 169, 174
Girioux, Dave, 61
Gladu, Roland, 106, 114, 118, 119, 120, 121
Globe, Sergeant, 46
'Goal Ball', 20
Godfrey, Paul, 13, 167
Goldie, John, 28, 31
Goldie, Thomas, 8, 31
Goldsmith, Fred, 32, 33, 39, 136, 159
Goliat, Mike, 166
Gomez, Lefty, 160
Goodhue, George, 17
Gorbous, Glenn, 72
Gorman, Harry, 32, 34, 37, 38, 39, 40
'Gossiper', the, 133, 134
Gott, Jim, 73
Grace, Paddy, 58
Grace, W.G., 133
Graham, Frank, 56
Graham, Jack, 46
Graham, Samuel, 21

Grant, Captain, 19
Grant, Frank, 143
Gray, Frank, 67, 119
Green, Marion (Hill), 60
Green, Wilson, 20
Gregg, Vean, 75
Gretzky, Wayne, 72, 124, 165
Grey Cup, 71
Grey, Romer, 30
Grey, Zane, 'The Winning Ball', 30
Griffey, Ken, 73
Griffin, Alfredo, 173, 176
Griffin, Don, 166
Griffith brothers (Bruce, Sherry, Billy, Jimmy), 126
Griffith, Calvin, 125, 156
Griffith, Clark, 125, 156
Griffith, Clark, II, 125
Griffith, Mildred, 125
Grissom, Marquis, 124
Groat, Dick, 164
Gove, Robert 'Lefty', 87
Gruber, Kelly, 174
Gruneau, Richard, 4
Guay, Donald, Introduction à l'histoire des sports au Québec, 20, 108, 112
Guillet, Maurice, 121
Gunns, the, 20
Guthrie, James, 43
Guzman, Juan, 174, 175, 176, 177, 179

Hacker, Rich, 179
Hael, Joseph, 108
Hahn, Fred, 166
Hall and McCulloch, 60 Years of Canadian Cricket, 131
Hallman, Bill, 47
Hamilton, Billy, 173
Hammonds, Jeffrey, 176
Handrahan, Vern, 55
Hanlan, Emily, 151
Hanlan, Ned, 151
Hansen, Ray, 48
Harkness, Tim, 82
Harnois, Arthur, 108
Harrington, Gordon, 49
Harris, Thomas, 94
Harrison, Mort, 42
Hartsfield, Roy, 171
Hartzell, Revd Howard, 92
Harvey, Doug, 118

Hastings, Kate, 152
Hastings, Scott, 37
Hatch, Benton, 67
Hatcher, Billy, 175
Haver, Jim 'Bunty', 28
Hawthorn, Tom, 104
Hawvermale, J.O., 57
Hayman, Arlo, 50
Haynes, Bruce, 125
Haynes, Joe, 125
Haywood, W.D., 95
Hearn, Bill, 28
Heas, C., 76
Heath, J. Geoffrey (Jeff), 160, 161
Helliwell, J., 131
Henderson, Rickey, 180, 181
Henke, Tom, 173, 175, 176, 177, 178
Henson, Josiah, 147
Hentgen, Pat, 179
Herman, Babe, 70
Herod, Dr, 29
Hess, Myrna (Hill), 60
Heward, J.O., 131
Hill, Dolly (Van Every), 60
Hill, Ruth (Van Every), 60
Hiller, John, 162
Hiraoka, Hiroshi, 99
Hirchner, John, 96
Hisle, Larry, 176
Hoak, Don, 117
hockey, 4, 9, 11, 14, 80, 113, 118, 139, 144, 155, 162, 164, 168
Holmes, John H., 131
Holmes, William 'Sweet Billy', 98
Hopper, Clay, 148
Hornung, Joe, 16, 33, 136
Horsman, Vince, 55
Houde, Camilien, 116
Hough, Charlie, 172
Howard, Elston, 166
Howarth, Jerry, 174
Howe, E.C., 44
Howe, Joseph, 28
Howell, Colin, 43, 45, 59, 60
Howley, Dan, 113, 162
Hudson's Bay Company, 19, 65, 73, 75, 84, 93
Hughes, Everett, 107
Humber, Frank, 58
Humber, William, 14
Hunt, Ron, 123

Hunter, London Tecumseh player, 1876, 33
Hunter, Bob, 11, 33
Hunter, Dorothy, 81
Hurby, Sergeant, 46
Hurst, Cam, 11
Hutcheson, St John, 131
Hyman, Ellis, 18

Imperial Oil Company, 37
Imperial Tobacco Company, 57
Imperial Trust, 169
Ireland, 21, 41, 133
Ito, Junji, 101

Jack, D.P., 22
Jackson, Darrin, 179
Jackson, Delores, 147
Jackson, Lane, 11
Jackson, Reggie, 55, 72
Jackson, Tommy, 48, 49
James, Bill, 13
Jamieson, Dorothy, 60
Japan, 1, 12, 98, 99, 100, 102, 106, 160
Jenkins, Ferguson, 147
Jenkins, Ferguson Arthur (Fergie), 82, 123, 147, 162–5
Jethroe, Sam, 115, 148, 166
Jewitt, Christine, 49
'Jim Crow' legislation, 143
Johnson, Arleene 'Johnnie', 80, 81
Johnson, Ben, 178
Johnson, Chappie, 118, 147
Johnson, Connie, 166
Johnson, Emil, 76
Johnson, Ollie, 144
Joliet, Aurel, 162
Jones, Bill, 30
Jones, Mel, 114
Jordan, Michael, 74, 180
Joss, Addie, 92
Judd, Oscar, 161
Jury, Jim, 32

Kahn, Roger, 13, 169
Karn, Chris, 17
Karn, Daniel, 16
Karn, Harry, 16
Karsay, Steve, 180
Kato, George, 101
Kaufman, Tony, 70

Kawashiri, Iwaichi, 101
Keating, Jack, 105
Keerl, George, 31
Kelly, Edward, 43
Kent, Jeff, 171, 176
Kent Cooke, Jack, 149, 151, 155
Kerr, Buddy, 166
Key, Jimmy, 107, *173*, *174*, 178, 179, 181
Kilkenny, Mike, 162
King, Dr, 133
King of the Hill, National Film Board documentary, 162
Kinsella, W.P., 59; *Shoeless Joe*, 75, 169
Kitchen, Mr, 38
Kitigawa, Eddie (Ed), 100, 101
Kitigawa, Hatsu (Mickey), 100
Kitigawa, Yo Horii, 100, 101
Kittle, Ron, 72
Knight, Joseph (Jonas), 159
Knorr, Randy, 179
Knowles, J.S., 22
knur and spell, 15
Korenaga, Abe, 102
Korince, George, 162
Kress, Charley, *166*
Kubek, Tony, 164
Ku Klux Klan, 102, 142–3

Labatt Breweries, 13, 168, 169
Labine, Clem, 109
Labour, Knights of, 98
LaCasse, Geoff, 94, 96, 98
Lachance, Alfred, 108
Lachance, Candy, 109
Lachappelle, Lucien, 118
Lacoursiere, Bob, 12
lacrosse, 3, 4, 5, 7, 8, 14, 84, 98, 109, 111, 133, 156
La Flamme, *76*
Lajoie, Napoleon 'Larry', 109
Lamb, Charlie, 104
Lambert, Thomas, 45, 46
Landeck, Arnie, *166*
Lanier, Max, 120, 121
Lannie, Joe, 156
La Rivière, Edmond Armand, 108
LaRoque, Sam, 108
Larwill, Al, 98
Lasorda, Tom (Tommy), *117*
Latham, George, 32, 33
Laurier, Prime Minister Wilfrid, 99

Lavigne, Paul, 13
Leduc, R.A., 118
Ledwith, *33*
Lee, Bill, 55
Lee, George 'Knotty', 156, 157
Lee, Manny, 178
Leitner, Irving, *Diamond in the Rough*, 1
Lemelin, Roger, *The Plouffe Family*, 107
Lent, L.B., 134
LePine, Pete, 108
Les Plouffes, 107
Levesque, Jean-Louis, 122
Leyva, Nick, 179
Lichtenhein, Sam, 113
Light, Colonel A.W., 17
Lightfoot, Charlie, 144
Linton, Doug, 175, 176
Little, Olive, *81*
Lizotte, Abel, 109
Lombardi, Vic, *166*
Lord's Cricket Ground, *10*
Loscombe, Mr, 136
Love, 25
Lowe, Ian, 89
Lund, Dr Bobby, 55
Lussier, Donat, 108

McAteer, 89
McCabe, Fred 'Husky', 56
McCann, 26, 27, 144
McCaskill, Kirk, 72
McClung, Nellie, 75, 77
McConochie, R., 133
McCrum, Joe, 96
McDonald, Allan, *46*
Macdonald, John A., 130
MacDonald, Jimmy 'Fiddler', 55
MacFarlane, Tom, 55
McGraw, John, 101
McGregor, Scott, 82
McGriff, Fred, *172*
McHale, John, *122*
MacIntyre, Manny, 54
Mack, Connie, 13, 91
Mack, Ed, 151, 155
McKay, Dave, 169, *170*
McKeever, 56
McKell, Wayne, 21
McKenzie, Ken, 162
McKenzie, Tom, *11*

Mackenzie, William Lyon, 4, 17, 99, 130
McKerlie, Almer, 88
McKillop, Bob, *11*
McLachlan, James Bryson, 49
McLain, Denny, 12
McLean, Larry, 45, 47
McLean, Stuart, *Welcome Home*, 108
Macleod, Starr, 49
MacMechan, Archibald, 9
McMurtry, W.J., 133
McNames, Nathaniel, 16
McQuaid, Vince 'Lefty', 52
McTavish, Malcolm, 133
McWhinnie, Robert, 25
Maddocks, Charlie, 28
Madison, 131
Magill, Charles, 23
Maglie, Sal 'Barber', 120, 121
Maksudian, Mike, 176, 177
Maldonado, Candy, 173, 177
Malloy, Jerry, 163
Mandeville, Pierre, 108
Mandeville, 'Snooks', 48
Mantle, Mickey, 121
Manush, Heinie, 70
Maracle, Olive, 60
Maranda, Georges, 106
March, 96
Marchildon, Phillipe (Phil), 162
Marsh, Lou, 163
Marston, Clay, 8
Martin, Billy, 29
Martin, Jean (Garlow), *60*
Martin, Lawrence, *Pledge of Allegiance*, 3
Martin, Robert, 75
Martin, Sam, 115
Martin, Vera, *60*
Meany, Tom, 139
Melilli, Beano, 87
Melillo, Oscar, 70
Mercer (Mercier), Win, 108
Meredith, Richard, 32, 139
Metcalfe, Alan, 8
Mexico, 89, 106
Middleton, General, 75
Midgely, John, 26
Miller, Hugh, 159
Miller, Vivian (Smith), *60*
Miller, W.E., *76*
Minarcin, Rudy, *166*

Miner, Bill, 65
Miner, Horace, 107
Mitchell, Fred, 45
Mitchell, Orville, 50
Mitchell, 'Shufflin' Bill, 48
Miyasaki, Harry, 100, 101
Miyata, Tom, 100, 101
Moffett, Samuel, 7
Molini, Albert, 121
Molitor, Paul, 176, 177, 178, 179, 180, 181
Molson Breweries, 74, 104
Monday, Rick, 124, 125
Monette, Judge Amadée, 118
Mooney, J., 111
Moore, Ed, 32, 34
Moore, Hazelton, 21
Morden, Dr, 32
Morrill, Simeon, 17
Morris, Jack, 171, 172, 174, 175, 176, 177, 181
Morrison, 25
Morton, Lew, 166
Moss, Howie, 120
Mott, Morris, 83, 85–6
Mountain Landis, Judge Kenesaw, 9, 113
Mulliniks, Rance, 172
Mulroney, Brian, 3
Munro, John, 13
Murphy, Dale, 172
Murphy, Mary Elizabeth, 109
Murphy, P., 158
Mussina, Mike, 179
Myers, Greg, 174
Myers, Hank, 31

Nagano, Manzo, 98
Nakamura, Frank, 101
Nakamura, Sally, 101
Nakanishi, Ken, 102
Nakatsu, Terry, 104
Native ball, 59–61
Nelson, Rocky, 117
Neville, Charles, 131
Newcombe, Don, 115
Nicol Fox, Helen, *81*
Nicholas, Dominic, *97*
Nichols, Jim, 28
Niekro, Phil, 55
Nimi, Joe, 100
Nishidera, Roy, 101
Nishizaki, Yosomatsu (Yo), 100

Nixon, Otis, 177
Noda, Ken, 102
Noguchi, Kunitaro, 99
Nomura, Dr Saitaro, 100
'northern spell', 15
North Western Coal and Navigation Company, 62
North West Company, 65
North-West Mounted Police, 62, 75, 77, 105
Nunn, Clyde, 49

Oakes, Maurice, *11*
O'Brien, John, 42
O'Dea, Art (Artie), 65, 89
O'Doul, Lefty, 160
O'Hara, Bill, 156
O'Hayer, 67
Oka, Yoshi, *100*
'old fashion' game, 22
Olerud, John, 170, 179, 180
Olivier, David, 108
Olympic code, 12
Olympic Games, 13, 160
O'Neill, Frank 'Tip', 44, 111, 113, 142
O'Neill, Harry, 159
O'Neill, James Edward 'Tip', *160*
O'Neill, Paul, 22, 56
O'Neill, Thomas P. 'Tip', Jr, 160
Ornest, Harry, 104
Orr, John B., *58*
Owen, Mickey, 120
Oxley, Henry, 55

Pagan, Dave, 82
Page, Joe, 11, 113, 118
Paige, Leroy 'Satchel', 83, 118, 149, 159
Pammett, Howard, *A History of the Township of Emily ...*, 9
Panciera, Guido, 49
Papke, Ernie, 100
Paradis, Alphonse, 108
Paradis, Jean-Marc, *100 Ans de Baseball à Trois-Rivières*, 108
Pardellian, J.B., 111
Parker, Dave, 172
Parker, Diane, 73, 74
Parker, Russ, 73, 74
Pascoe, 26
Pascoes, the, 20
Pasquel, Jorge, 120
Passmore, William, 80

Pattison, Jimmy, 104
Payette, F., 112
Payzant, Joan, 21–2
Pearson, Prime Minister Lester B., 13, *122*, 165
Pelissier, Tim, *76*
Pellatt, Sir Henry, 10
Peloquin, Hector, 108
Perkins, Dave, 181
Perrin, Daniel, 32
Perry, Mattie (French Mattie), 43
Peters, William, 20
Piche, Ron, 106
Pickering, William, 42, *46*
Pillsbury, Corinne Griffith, 125
Piniella, Lou, 73
Plamondon, J.P., 112
Playter, Ely, 15
Pocklington, Peter, 72, 75
Podres, Johnny, 117
Pollock, Sam, 168
Powers, *33*
Prentice, Bobby, 179
Puhl, Terry, 82

Quantrill, Paul, 175
Quisenberry, Dan, 162

rabbit ball, 30
Racine, Hector, 113, 114
Rafferty, 104, 134
Railton, George, 32
Railways: Canadian Northern, 65; Grand Trunk, 38; Great Western, 23, 24, 27, 38; Intercolonial, 46; London, Huron and Bruce, 18; Temiskaming and Northern, 37
Raines, Tim, 123
Randall, Mr Joseph, 15
Rathburne, 96
Rattlesnake, Jimmy, 59
Raymond, Claude, 106, 123
Reardon, Beans, 70
Reid, Sir Robert, 56
Reid-Newfoundland Company, 57
Richard, Maurice, 118
Richardson, Ernie, 82
Richler, Mordecai, 113
Ricker, Owen, 67
Rickey, Branch, 113, 114, 115, 120, 121, 139, 162
Riel, Louis, 50, 75, 83
Risberg, Swede, *88*

Ritchie, Grace, 45
Ritchie, Sam, 111
Robb, Douglas 'Scotty', 50
Roberts, Robin, 164
Robertson, Al, *11*
Robertson (later Clark), Calvin, 125
Robertson, Jane, 125
Robertson, Jimmy, 125
Robertson (later Clark), Thelma, 125
Robichaud, Premier Louis, 13
Robinson, Cliff, 89
Robinson, Jackie, 29, 107, 114–15, 120, 121, 139, *140*
Robinson, Laura, 60
Robinson, Wilbert, *70*
Rodriguez, Hector, *166*
Rogers, Steve, 72, 125, *124*
'ronde', 15
Ross, Alex, 143
'Round Ball', 20
rounders, 15, 19, 20, 22
Rowe, Bill, 112, 113
rowing, 104, 155
Roy, Amadée, 118
Roy, Jean Pierre, 114, 120, 121
Russell, Burton, 11
Ruth, Babe, 13, 102, 154, 165
Rutledge, Peter, 56
Rykert, C.J., 131

Salas, Wilfredo, *119*
Salmon, Tim, 72
Samuells, A.R., 27
Sato, Mickey, *101*
Savard, Ernest, 113
Sawamura, Eiji, 102
Scarff brothers, 28: Jim, 27; Tom, 31
Schaeffer, Harry, *166*
Schofield, historian of Manitoba, 19
Schofield, Dick, 72, 179, 181
Schofield, Jim, 136, 137
Schooler, Mike, 173
Scott, Gladwyn, *11*
Scott, Glennis, *11*
Scott, Walter, 75, *76*, 77
Scriven, George, *46*
Scriven, W.R., *46*
Seafoot, Cliff, *11*
Seaman, Danny, 54
Seaman, Garneau, 54, 55
Selkirk, Lord, 19

Sellar, Robert, 21
Settlemire, Merle 'Lefty', 50
Sevigny, J.E., 112
Sewell, Luke, *166*
Sexton, F.J., 44
Shallow, Mike, *58*
Sharperson, Mike, 175
Shaughnessy, Frank 'Shag', 113, 114, 115
Shaw, Thomas (Tom), 133, 135, 136
Shepperd, George, 83
shinny, 19
Shirley, Tex, 121
Shirriff, Dr, 21
Shore, Ray, 166
Shuba, George 'Shotgun', 117
Shury, David, 12, 13, 75, 78, 82, 149
Shuttleworth, Bill, 23, 26
Shuttleworth, Jim, 25, 26
Siddall, Joe, 107, 158
Simcoe, Lieutenant-Governor, 17
Sinclair, May, 61
Sinclair, Sam, 61
Sinclair, Walter, 61
Skalbania, Nelson, 104
Sleeman Breweries, 30
Sleeman, George, 5, *30*, 31, 35, 36, 37, 144
Sleeman, John, 30
Smiley, Bob, 23
Smith, Bill, 29, 31, *166*
Smith, Eldorado, 105
Smith, Goldwin, 3
Smith, Howard, *46*
Smith, Larry, *11*
Smith, Lonnie, 177
Smith, Ron, *11*
Smith, Tommy, 39
Smith, William, *46*
Snider, Duke, 55
snowshoe teams, 111
Snyder, Gerald, 123
Sockalexis, Louis 'Chief', 59
softball, 8, 10, 54, 60, 77, 79, 80, 81, 162
Solman, Lol, 151, 155
Somerville, Ed, 39
Sonneborn, Solomon, 37
South Africa, 46
Southam, Richard, 32, 38
Southam, William, 7, 8, 32, 38
Spalding, *Canadian Base Ball Guide/Official Guide*, 7, 8, 50

Spence, Harry, 30
Sporting News, 67, 89, 122, 168–9
Sports Illustrated, 164
Sprague, Ed, 175, 176, 177, 180
Squair, John, 130, 132
Squarebriggs, Johnny 'Snags', 53
Squire Hill, Kay, *60*
Squire Hill, Mel, *60*
Starkman, Howard, 178, 180, 181
Starffin, Victor, 102
Staub, Rusty 'le grand orange', *123*
Stead, Ron, *11*, 12
Steinbrenner, George, 172
Stengel, Casey, 174
Stevens, Ed, *166*
Stevenson, Ephraim (Eph), 28, 29
Stewart, Dave, 177, 179, 180
Stewart, Don, 101
Stieb, Dave, 172, 174, 175, 177
Stone, Ross, *11*
Stoneman, Bill, 123
stoolball, 15
Stottlemyre, Todd, 173, 180
Stovey, George, 143
Stowe, Harriet Beecher, 147
Stuart, Herb, 89
Suga, Kenichi 'Ty', 101, *102*
Sullivan, John, 9, 181
Summer, Don, *11*
Sutcliffe, Rick, 175
Sutton, Thomas C., 131
Sweeney, Ed, 90, *91*, 92

Tamblyn, William Ware, 132
Tanaka, Charlie, 101
Tanner, Ed, *11*
Tansey, 'Lil Abner', 48
Tappe, Ted, 72
Tarbell, Joe, 59
Tartabull, Danny, 74
Taylor, Russ, 123
'tec, thèque, la grande thèque', 15, 21
Teitel, Jay, 163
Templeton, John, 132, 134
Thayer, Ernest Lawrence, 159
Theakston, R.S., 48
Thompson, Ryan, 176
Thurrier, Alfred, 118, 120
Thurston, Earl, 69
Timlin, Mike, 177, 181

Tiptoe, Don, 89
townball, 15, 18, 19
trading card, Zee-Nut series, *142*, 144
trapball, 15
Trouppe, Quincy, 83, 121
Trudeau, Charles, 113, 114
Trudeau, Pierre Elliott, 113, 114, 165
Tulk, A.E., 98, 102
Turk, Frank, 93, 94
Turner, J. Lloyd, *65*
Tuttle, Annie, 43

Uchiyama, Yuji, 100
Ueberroth, Peter, 74
United States, 1, 4, 9, 10, 12, 13, 17, 18, 19, 21, 23, 24, 27, 38, 70, 83, 108, 132, 141, 142, 155, 156, 160, 163
    Alabama, 55, 180
    Alaska, 61, 105
    California, 9, 27, 65, 70, 71, 72, 73, 74, 89, 93, 94, 139; Angels 'Halos' 72, 74, 174; Athletics, 65; Buffalos, 71; Cannons, 73, 74; Glendale, 70; L.A. Dodgers, 58, 72, 117, 123, 124, 161, 175; Oakland, 142, 144, 169, 171, 176, 177, 178, 180; Odeons, 73; Purity 99 team, 71; Sacramento, 161; San Diego, 122, 172, 179; San Francisco, 93, 95, 106, 160, 169, 177; San Jose, 93; Santa Clara, 93; USC team, 72
    Colorado, 16, 91, 92, 95
    Dakotas, 4, 67
    District of Columbia: Washington Senators, 108, 119, 125, 126, 156, 159, 161, 165
    Florida, 113, 171; Homestead Grays, 149; Marlins, 73, 178; Senior Professional B Assocation, 165; West Palm Beach, 123
    Georgia, 77; Atlanta, 39, 90, 134, 165, 177
    Hawaii, 93, 99, 102, 106
    Idaho, 72, 160
    Illinois, 17; Chicago, 26, 29, 35, 46, 71, 80: Cubs, 39, 79, 123, 163, 164, 168, 177, 179, 180, Ladies, 45, White Stockings/White Sox, 39, 87, 89, 124, 173; Evanston: Northwestern University, 65, 67; Peoria Redwings, 80; Rockford, 29
    Indiana, 17; Fort Wayne Daisies, 81; Indianapolis, 111, 146, 148, 149; South Bend, 80
    Iowa: Cedar Rapids, 90
    Kentucky: Louisville, 37, 38, 39, 50, 106, 115

Louisiana: New Orleans, 134

Maine, 4, 109; Bangor, 42; Hancock, 108; Lewiston, 42; Milltown, 50; New Brunswick-Maine pro league, 45; Penobscot, 59; Waterville, 45; Winthrop, 109

Maryland: Baltimore, 29, 82, 108, 115, 159, 165, 176

Massachusetts, 4; Boston, 31, 41, 109, 159: American League team, 45, Bloomers women's team, 63, 77, 98, Braves, 50, 106, 119, 161, Harvard University, 44, John P. Lovell Arms Company, 46, Marathon, 172, Red Sox/Red Stockings, 29, 30, 38, 39, 50, 55, 121, 134, 156, 161, 173, 175, Woven Hose team, 46; Lowell, 108; Lynn, 36; Marlboro, 108; Southbridge, 108; Webster, 108; Worcester, 39, 108, 159

Michigan, 4, 5, 150; Bay City, 159; Detroit, 5, 17, 23, 27, 159, 162, 172, 176, 179: Brother Jonathon Club, 27, Cass Club, 138, Rialtos black team, 142, Tigers, 12, 159, 162, 163; Jackson Unknowns, 28; Kalamazoo Lassies, 80; North Michigan and Michigan State League, 143; Port Huron, 27; River Rouge Ford plant, 159

Minnesota, 4, 67, 106, 108, 176; Duluth, 89; Minneapolis, 90, 92, 96, 159; Minneapolis-St Paul, 83, 126; Minneapolis Twins, 125, 171, 173; University of Minnesota, 90

Missouri: Kansas City, 55, 73, 82, 162, 169, 178: Athletics, 55, Monarchs, 83, 89, 149, Royals, 178; St Louis, 36, 67, 86: Black Sox, 142, 144, Browns, 113, 119, 142, 159, 160, 161, Cardinals, 12, 50, 73, 124, 177

Montana, 62, 87

Nebraska, 72

New Hampshire, 4, 157

New Jersey, 50, 157; Jersey City, 91, 115, 143; Newark, 92, 113, 115, 143, 148, 167

New Mexico, 55

New York, 4, 9, 14, 21, 132, 139, 156; Buffalo, 39, 70, 92, 111, 113, 122, 143: Cuban Giants, 142, 144, 149, Eastern League team, 31, Niagaras, 25; Cooperstown, 1, 2, 25, 134, 139, 163; Herkimer County, 25; New York City, 18, 19, 22, 31, 35, 36, 37, 109, 131, 174, 176, 179: American League team, 91, Brooklyn Atlantics, 25, 27, Brooklyn Dodgers, 13, 49, 54, 70, 109, 113, 114, 115, 117, 120, 121, 139, Excelsiors, 25, Fly

Away team, 135, Giants, 108, 121, 156, Highlanders (Yankees), 91, 92, Knickerbockers, 18, 22, 23, 93, Manhattan, 130, Mets, 82, 124, 156, 162, 164, 172, 176, Yankees, 82, 113, 121, 153, 165, 172, 179; Ogdensburg, 21; Oneida, 30, 31, 142; Oneonta, 161; Oswego, 19; Poughkeepsie: Vassar College teams, 77; Rochester, 23, 27, 29, 39, 50, 89, 117, 132, 133, 143, 161; Syracuse, 36, 39, 113, 115, 143, 159, 175, 176; Watertown, 26, 30, 135, 142; Wellsville, 55

North Carolina, 120; Charlotte, 125; Rockingham, 90, 92

North Dakota, 87, 88, 118; Bismarck, 83; Enderline, 90; Fargo, 89; Minot, 89; Williston, 89

Ohio, 4, 9, 17, 118, 179; Akron, 158; Cincinnati, 72, 109, 111, 138; Cleveland, 29, 37, 45, 92, 107, 161; Columbus, 161; Toledo, 92

Oklahoma, 67, 141

Oregon, 72, 96; Portland, 98

Pennsylvania 4, 9, 150, 159, 161, 162; Philadelphia 4, 16, 35, 36, 45, 77, 179: Athletics, 45, 47, 91, 160, 162, National League team, 114, Phillies, 124, 164, 180, 181; Pittsburgh, 27, 36, 38, 39, 109: Alleghenies, 39, Pirates, 121, 123, 176, 181; Williamsport, 143

Rhode Island, 44, 109; Manville, 108; Pawtucket, 50; Providence, 39, 165

South Carolina: Myrtle Beach, 176, 179

Tennessee: Chattanooga, 125

Texas, 9, 179; Corsicana team, 158; Houston Astros, 82; Rangers, 125, 164

Utah: Ogden A's, 72; Salt Lake City, 73, 74, 160

Vermont, 4, 17

Washington, 70, 72, 125; Fort Lewiston, 71; Port Townsend, 96; Seattle, 98, 100, 112, 160, 169, 175: Mariners, 73, 74, 173, 179, Rainiers, 161; Spokane, 117; Tacoma, 98, 144

West Virginia, 108

Wisconsin: Eau Claire, 55; Green Bay, 39; Kenosha Comets, 79; Lacrosse, 67; Milwaukee, 169, 172, 181: Braves, 55, Brewers 'Brew Crew', 176, 179, 180; Superior, 89

Upham, John, 158

Vadeboncoeur, Onesime Eugene, 106, 108

Valenzuela, Fernando, 72

Van Horne, Sir William, 98
Vaniderstine, Ches, 52
Vizquel, Omar, 74
Von Der Ahe, Chris, 142

Waddington, Alfred, 93
Wagner, Honus, 1
Waite, L.C., 36
Walker, Fleetwood, 143
Walker, Larry, 124
Walker, Moses, 143
Ward, Duane, 173, 176, 177, 179, 181
Washington, U.L., 55
Waters, Evelyn, 8
Wawryshyn, Evelyn 'Evie', *81*
Webber, 45
Webster, Mitch, 168
Weadick, Guy, 73
Weisler, Bob, *166*
Weiss, William, 142
Wells, David, 174
Werry, Benjamin (Ben), 127, 132, 136
Westbrook, Andrew, 17
Whelly, Jim, 44
White, William Freemont 'Deacon', 65, 67, 69, 71
White, Devon, 72, 172, 177, 179, 181
Whitmore, Darrell, 73
Whitson, David, 4
Whittle, Chuck, 50
Wilcox, Jim, 138
Wilkes, Jimmy, *146*, 148
Wilkie, Aldon, 176
Williams, Kathy, 147

Williams, Mitch 'Wild Thing', 180
Williams, Ted, 120, 179
Wilson, Alfred, 118
Wilson, Archie, *166*
Wilson, Nigel, 73, 178
Wilson, 'Rube', 48
Wiman, Erastus, 155, 165
Winfield, Dave 'Mr May', 171, 172, 173, 176, 177, 178, 179
Wingo, Ed, 108, 118
Winks, Robin, 139
Wise, Rick, 55
Wolfe, Frank, 71
Wolseley, Colonel, 83
Wood, Charles, 55
Wright, Richard, *Black Boy*, 139
Wright, W., 46
Wright, Harry, *134*
Wrigley, Phil, 79, 80
Wyld, Richard, 75
Wyld, Robert, 75

Yamamura, Roy, 101
Yasui, Reggie, 101
Yoshinaka, George, 102
Yoshioka, Harry, *100*
Young, Cy, 91, 163
Young, Joe, 14
Yount, Robin, 172

Zabala, Adrian, *119*, 120, 121
Zamloch, Carl, *65*
Zeman, Brenda, 59